Collins

AQA GCSE (9-1)
Physics
Grade 8/9 Booster Workbook

Lynn Pharaoh

William Collins' dream of knowledge for all began with the publication of his first book in 1819. A self-educated mill worker, he not only enriched millions of lives, but also founded a flourishing publishing house. Today, staying true to this spirit, Collins books are packed with inspiration, innovation and practical expertise. They place you at the centre of a world of possibility and give you exactly what you need to explore it.

Collins. Freedom to teach

HarperCollins Publishers
The News Building
1 London Bridge Street
London SE1 9GF

Browse the complete Collins catalogue at www.collins.co.uk

First edition 2016

10 9 8 7 6 5 4 3

ISBN 978-0-00-819435-2

www.collins.co.uk

A catalogue record for this book is available from the British Library

Commissioned by Joanna Ramsay
Project managed by Sarah Thomas and Siobhan Brown
Copy edited by Lynette Woodward
Proofread by David Hemsley
Answer check by Tony Clappison
Typeset by Jouve India Pvt Ltd.,
Artwork by Jouve India Pvt Ltd.
Cover design by We are Laura and Jouve
Cover image: Shutterstock/Triff for the bottom photo and Spyros/Shutterstock for the notepad
Printed by Grafica Veneta S.p.A., Italy

Contents

Section 1 • Energy changes in a system

Section 2 • Electricity

Section 3 • Particle model of matter

Section 4 • Atomic structure

Section 5 • Forces

Section 6 • Waves

Section 7 • Magnetism and electromagnetism

Section 8 • Space physics

Answers

Introduction

This workbook will help you build your confidence in answering Physics questions for GCSE Physics and GCSE Combined Science.

It gives you practice in using key scientific words, writing longer answers, answering synoptic questions as well as applying knowledge and analysing information.

You will find all the different question types in the workbook so you can get plenty of practice in providing short and long answers.

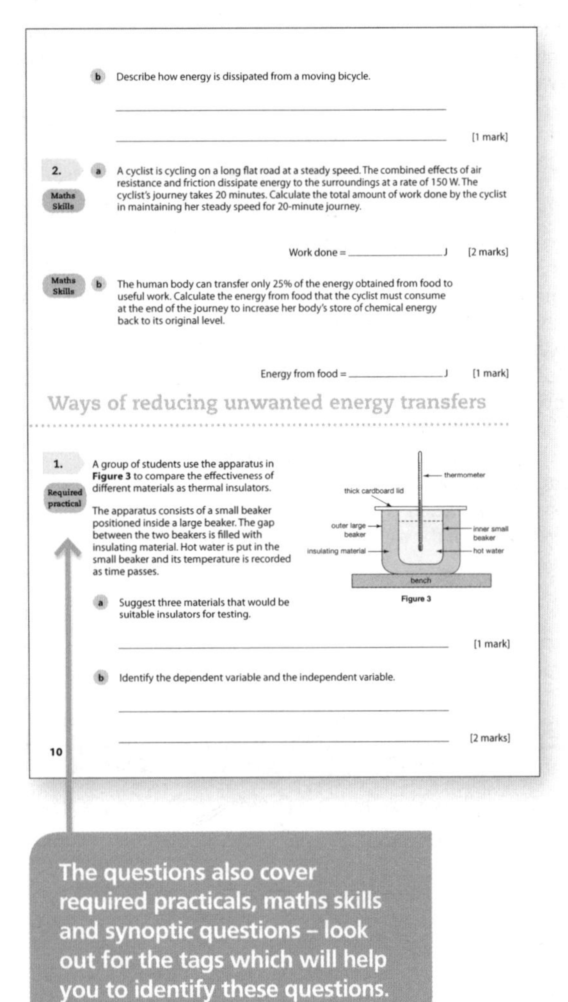

b Describe how energy is dissipated from a moving bicycle.

____

____________________ [1 mark]

2. Maths Skills **a** A cyclist is cycling on a long flat road at a steady speed. The combined effects of air resistance and friction dissipate energy to the surroundings at a rate of 150 W. The cyclist's journey takes 20 minutes. Calculate the total amount of work done by the cyclist in maintaining her steady speed for 20-minute journey.

Work done = __________ J [2 marks]

Maths Skills **b** The human body can transfer only 25% of the energy obtained from food to useful work. Calculate the energy from food that the cyclist must consume at the end of the journey to increase her body's store of chemical energy back to its original level.

Energy from food = __________ J [1 mark]

Ways of reducing unwanted energy transfers

1. Required practical A group of students use the apparatus in **Figure 3** to compare the effectiveness of different materials as thermal insulators.

The apparatus consists of a small beaker positioned inside a large beaker. The gap between the two beakers is filled with insulating material. Hot water is put in the small beaker and its temperature is recorded as time passes.

Figure 3

a Suggest three materials that would be suitable insulators for testing.

____________________ [1 mark]

b Identify the dependent variable and the independent variable.

____

____________________ [2 marks]

10

The questions also cover required practicals, maths skills and synoptic questions – look out for the tags which will help you to identify these questions.

Learn how to answer test questions with annotated worked examples.

This will help you develop the skills you need to answer questions.

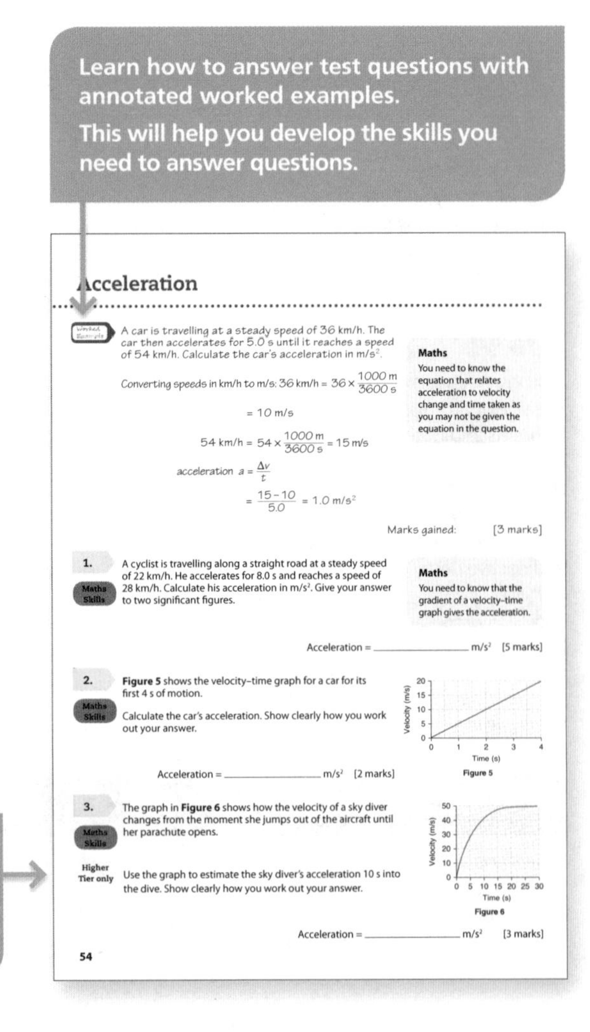

Acceleration

Worked Example A car is travelling at a steady speed of 36 km/h. The car then accelerates for 5.0 s until it reaches a speed of 54 km/h. Calculate the car's acceleration in m/s².

Converting speeds in km/h to m/s: $36\text{ km/h} = 36 \times \frac{1000\text{ m}}{3600\text{ s}}$

$= 10\text{ m/s}$

$54\text{ km/h} = 54 \times \frac{1000\text{ m}}{3600\text{ s}} = 15\text{ m/s}$

acceleration $a = \frac{\Delta v}{t}$

$= \frac{15 - 10}{5.0} = 1.0\text{ m/s}^2$

Marks gained: [3 marks]

Maths You need to know the equation that relates acceleration to velocity change and time taken as you may not be given the equation in the question.

1. Maths Skills A cyclist is travelling along a straight road at a steady speed of 22 km/h. He accelerates for 8.0 s and reaches a speed of 28 km/h. Calculate his acceleration in m/s². Give your answer to two significant figures.

Maths You need to know that the gradient of a velocity-time graph gives the acceleration.

Acceleration = __________ m/s² [5 marks]

2. Maths Skills **Figure 5** shows the velocity–time graph for a car for its first 4 s of motion.

Calculate the car's acceleration. Show clearly how you work out your answer.

Acceleration = __________ m/s² [2 marks]

Figure 5

3. Maths Skills The graph in **Figure 6** shows how the velocity of a sky diver changes from the moment she jumps out of the aircraft until her parachute opens.

Higher Tier only Use the graph to estimate the sky diver's acceleration 10 s into the dive. Show clearly how you work out your answer.

Acceleration = __________ m/s² [3 marks]

Figure 6

54

Higher Tier content is clearly marked throughout.

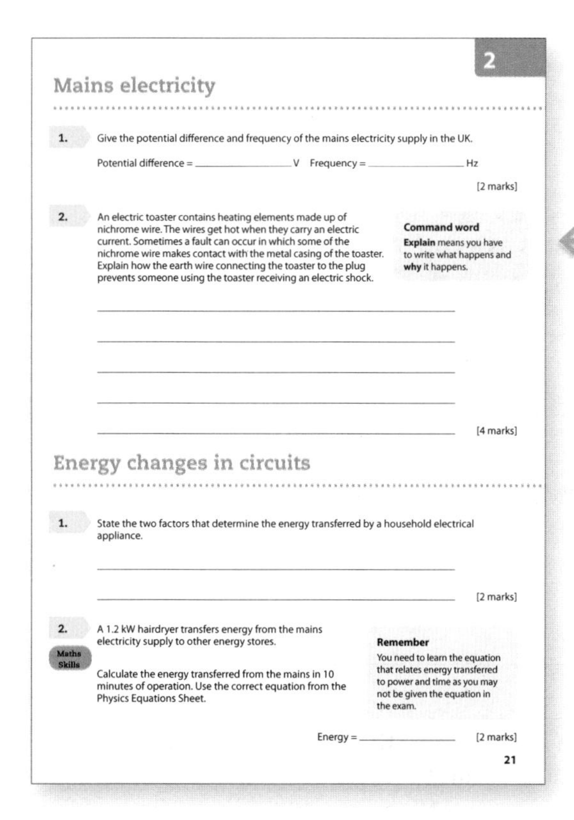

2

Mains electricity

1. Give the potential difference and frequency of the mains electricity supply in the UK.

Potential difference = ________ V Frequency = ________ Hz

[2 marks]

2. An electric toaster contains heating elements made up of nichrome wire. The wires get hot when they carry an electric current. Sometimes a fault can occur in which some of the nichrome wire makes contact with the metal casing of the toaster. Explain how the earth wire connecting the toaster to the plug prevents someone using the toaster receiving an electric shock.

Command word
Explain means you have to write what happens and **why** it happens.

[4 marks]

Energy changes in circuits

1. State the two factors that determine the energy transferred by a household electrical appliance.

[2 marks]

2. A 1.2 kW hairdryer transfers energy from the mains electricity supply to other energy stores.

Maths Skills

Calculate the energy transferred from the mains in 10 minutes of operation. Use the correct equation from the Physics Equations Sheet.

Remember
You need to learn the equation that relates energy transferred to power and time as you may not be given the equation in the exam.

Energy = ________ [2 marks]

21

There are lots of hints and tips to help you out. Look out for tips on how to decode command words, key tips for required practicals and maths skills, and common misconceptions.

The amount of support gradually decreases throughout the workbook. As you build your skills you should be able to complete more of the questions yourself.

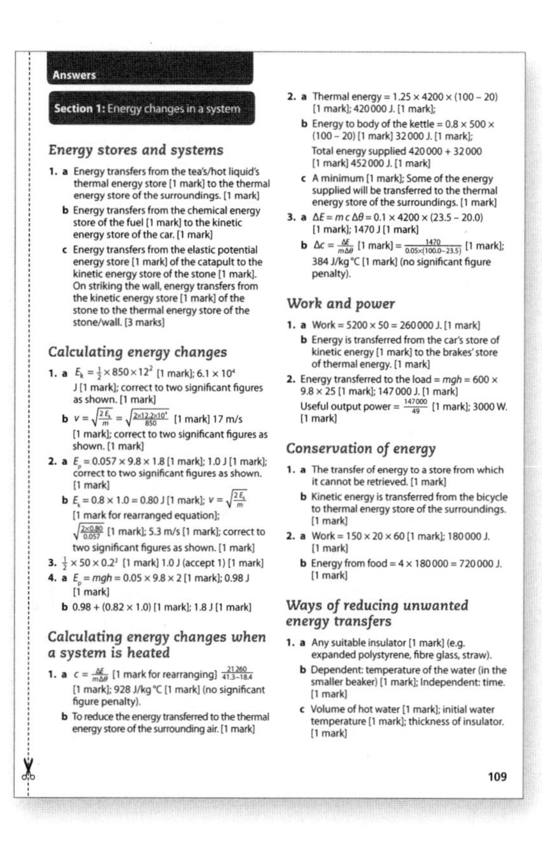

Answers

Section 1: Energy changes in a system

Energy stores and systems

1. a Energy transfers from the tea's/hot liquid's thermal energy store [1 mark] to the thermal energy store of the surroundings. [1 mark]
   b Energy transfers from the chemical energy store of the fuel [1 mark] to the kinetic energy store of the car. [1 mark]
   c Energy transfers from the elastic potential energy store [1 mark] of the catapult to the kinetic energy store of the stone [1 mark]. On striking the wall, energy transfers from the kinetic energy store [1 mark] of the stone to the thermal energy store of the stone/wall. [3 marks]

Calculating energy changes

1. a $E_k = \frac{1}{2} \times 850 \times 12^2$ [1 mark]; 6.1×10^4 J [1 mark]; correct to two significant figures as shown. [1 mark]
   b $v = \sqrt{\frac{2E_k}{m}} = \sqrt{\frac{2 \times 12.2 \times 10^4}{850}}$ [1 mark] 17 m/s [1 mark]; correct to two significant figures as shown. [1 mark]
2. a $E_p = 0.057 \times 9.8 \times 1.8$ [1 mark]; 1.0 J [1 mark]; correct to two significant figures as shown. [1 mark]
   b $E_k = 0.8 \times 1.0 = 0.80$ J [1 mark]; $v = \sqrt{\frac{2E_k}{m}}$ [1 mark for rearranged equation]; $\sqrt{\frac{2 \times 0.80}{0.057}}$ [1 mark]; 5.3 m/s [1 mark]; correct to two significant figures as shown. [1 mark]
3. $\frac{1}{2} \times 50 \times 0.2^2$ [1 mark] 1.0 J (accept 1) [1 mark]
4. a $E_p = mgh = 0.05 \times 9.8 \times 2$ [1 mark]; 0.98 J [1 mark]
   b $0.98 + (0.82 \times 1.0)$ [1 mark]; 1.8 J [1 mark]

Calculating energy changes when a system is heated

1. a $c = \frac{\Delta E}{m \Delta \theta}$ [1 mark for rearranging] $\frac{21\,260}{41.3 - 18.4}$ [1 mark]; 928 J/kg °C [1 mark] (no significant figure penalty).
   b To reduce the energy transferred to the thermal energy store of the surrounding air. [1 mark]
2. a Thermal energy $= 1.25 \times 4200 \times (100 - 20)$ [1 mark]; 420 000 J. [1 mark];
   b Energy to body of the kettle $= 0.8 \times 500 \times (100 - 20)$ [1 mark] 32 000 J. [1 mark]; Total energy supplied 420 000 + 32 000 [1 mark] 452 000 J. [1 mark]
   c A minimum [1 mark]; Some of the energy supplied will be transferred to the thermal energy store of the surroundings. [1 mark]
3. a $\Delta E = m c \Delta\theta = 0.1 \times 4200 \times (23.5 - 20.0)$ [1 mark]; 1470 J [1 mark]
   b $\Delta c = \frac{\Delta E}{m \Delta \theta}$ [1 mark] $= \frac{1470}{0.05 \times (100.0 - 23.5)}$ [1 mark]; 384 J/kg °C [1 mark] (no significant figure penalty).

Work and power

1. a Work $= 5200 \times 50 = 260\,000$ J. [1 mark]
   b Energy is transferred from the car's store of kinetic energy [1 mark] to the brakes' store of thermal energy. [1 mark]
2. Energy transferred to the load $= mgh = 600 \times 9.8 \times 25$ [1 mark]; 147 000 J. [1 mark] Useful output power $= \frac{147000}{49}$ [1 mark]; 3000 W. [1 mark]

Conservation of energy

1. a The transfer of energy to a store from which it cannot be retrieved. [1 mark]
   b Kinetic energy is transferred from the bicycle to thermal energy store of the surroundings. [1 mark]
2. a Work $= 150 \times 20 \times 60$ [1 mark]; 180 000 J. [1 mark]
   b Energy from food $= 4 \times 180\,000 = 720\,000$ J. [1 mark]

Ways of reducing unwanted energy transfers

1. a Any suitable insulator [1 mark] (e.g. expanded polystyrene, fibre glass, straw).
   b Dependent: temperature of the water (in the smaller beaker) [1 mark]; Independent: time. [1 mark]
   c Volume of hot water [1 mark]; initial water temperature [1 mark]; thickness of insulator. [1 mark]

109

There are answers to all the questions at the back of the book. You can check your answers yourself or your teacher might tear them out and give them to you later to mark your work.

Energy stores and systems

1. **a** Describe the transfer of energy between stores as a cup of hot tea cools.

____

______________________________ [2 marks]

b Describe the transfer of energy between stores as a car accelerates from rest.

____

______________________________ [2 marks]

c A catapult is stretched and released to fire a stone, which then hits a wall.
Describe the transfer of energy between stores as time passes.

____

____

____

______________________________ [3 marks]

Calculating energy changes

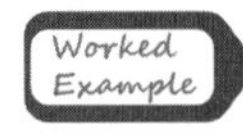

A moving car, of mass 850 kg, has a kinetic energy store of 110 000 J.

Calculate the speed of the car.

Kinetic energy: $E_k = \frac{1}{2}mv^2$ can be rearranged to give: $v = \sqrt{\frac{2E_k}{m}}$

So $v = \sqrt{\frac{2 \times 110\,000}{850}}$

$= 16$ m/s

Marks gained: [3 marks]

Maths

To round a number to two significant figures, look at the third digit. Round up if the digit is 5 or more. Round down if the digit is 4 or less.

1. Maths Skills

a A car of mass 850 kg is travelling at a constant speed of 12 m/s. Calculate the car's store of kinetic energy. Give your answer in standard form expressed to two significant figures.

Kinetic energy = ________________ J [3 marks]

Maths

Answers given in standard form should have only one digit in front of the decimal point. For example, 750 000 should be written 7.5×10^5.

b The car now accelerates increasing its store of kinetic energy by a factor of 2. Calculate the car's speed. Give your answer to two significant figures.

Speed = ________________ m/s [3 marks]

2. Maths Skills

a A tennis ball of mass of 57 g is lifted from the ground to a height of 1.8 m. Calculate the increase in the ball's store of gravitational potential energy. Give your answer to two significant figures.

Gravitational field strength = 9.8 N/kg.

Gravitational potential energy = ________________ J [3 marks]

b The tennis ball is released and falls to the ground. The work done by air resistance transfers 20% of the ball's store of gravitational potential energy to the surrounding air's store of thermal energy. Calculate the speed of the ball as it hits the ground. Give your answer to two significant figures.

Speed on hitting the ground = ________________ m/s

[5 marks]

Problem solving

In a 4- or 5-mark calculation question, there is usually more than one problem to solve to get the information needed to calculate the final answer. Always show your working. You may still get some marks even if the final answer is wrong.

3. Maths Skills

The length of an unstretched spring is 2.2 cm. Its spring constant is 50 N/m.
The spring is now stretched so that its length is 22.2 cm.

Calculate the spring's store of elastic potential energy.

Elastic potential energy = ________________ J [2 marks]

Remember

The equation for elastic potential energy, $E_e = \frac{1}{2}ke^2$, it can also be used to calculate the energy stored in a compressed spring. In this case, it represents the decrease in length of the spring.

4. Maths Skills

a. A stone of mass 50 g is fired vertically upwards using a catapult. The stone was 2.0 m above the ground at its release point. Calculate the stone's store of gravitational potential energy at its release point.

Gravitational field strength = 9.8 N/kg.

Gravitational potential energy = ____________ J [2 marks]

Maths Skills

b. There is 1.0 J of energy in the stretched catapult's store of elastic potential energy. 82% of this elastic potential energy is transferred to the stone's gravitational potential energy store as it reaches its maximum height. Calculate the total gravitational potential energy of the stone at its maximum height above the ground.

Gravitational potential energy = ____________ J [2 marks]

Calculating energy changes when a system is heated

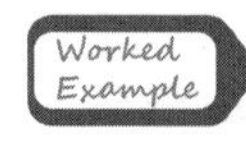

52 000 J of energy is transferred to the thermal energy store of 200 g of water.

Calculate the resulting rise in the water's temperature.

Specific heat capacity of water = 4200 J/kg/ °C.

$\Delta E = mc\Delta\theta$ can be rearranged to give the temperature rise, $\Delta\theta$:

$$\Delta\theta = \frac{\Delta E}{mc}$$

$$= \frac{52000}{0.2 \times 4200} = 62\text{ °C}$$

Marks gained: [3 marks]

Remember

Calculations involving materials getting hotter or cooler can be done using the following equation, which is on the Physics Equation Sheet:

$\Delta E = mc\Delta\theta$

You should be able to select and apply the equation – it may not be given in the question.

1. Required practical

A student uses the apparatus shown in **Figure 1** to measure the specific heat capacity of aluminium. A heater slots into the larger hole in the aluminium block and thermometer slots in the smaller hole. The aluminium block is wrapped in insulation such as expanded polystyrene.

The joulemeter records the amount of energy transferred to the thermal energy store of the heater. Energy is transferred from the heater to the aluminium block. The student switches the power supply unit on for 600 s and records the measurements (**Table 1**).

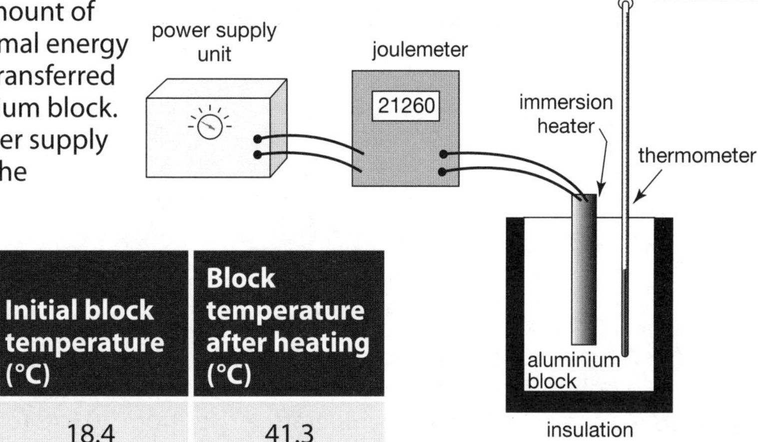

Figure 1

Mass of block (kg)	Joulemeter reading (J)	Initial block temperature (°C)	Block temperature after heating (°C)
1.00 kg	21260	18.4	41.3

Table 1

Maths Skills

a Calculate the specific heat capacity of aluminium. Choose the correct equation from the Physics Equation Sheet.

Specific heat capacity = ________________ J/kg °C [3 marks]

b Describe the purpose of the insulation wrapped around the aluminium block.

____

____ [1 mark]

2. Maths Skills

a An electric kettle contains 1.25 kg of water at 20 °C. The kettle is switched on and the water is heated to 100 °C. Calculate the increase in the thermal energy store of the water. The specific heat capacity of water is 4200 J/kg/ °C.

Increase in thermal energy = ________________ J [2 marks]

Maths Skills

b The kettle is made of stainless steel and has a mass of 0.80 kg. During the heating of the water, the kettle experiences the same temperature rise. Calculate the total amount of energy that must be transferred from the mains to enable the kettle to heat the water to 100 °C. Specific heat capacity of stainless steel = 500 J/kg °C.

Total energy transferred = ________________ J [3 marks]

c State whether your answer to Question 3(b) is likely to be a maximum or minimum value.

Explain your answer.

____

____ [2 marks]

3. Required practical

A brass mass of 50 g is heated to 100.0 °C. It is then lowered into a polystyrene cup containing 100 g of water (**Figure 2**).
As a result, the water temperature rises from 20.0 °C to 23.5 °C.

Specific heat capacity of water = 4200 J/kg °C.

Figure 2

a Calculate the thermal energy transferred to the water.

Thermal energy transferred = ________________ J [2 marks]

b Assume that no energy was transferred to the surroundings and that the water and the brass have the same final temperature. Calculate the specific heat capacity of brass.

Specific heat capacity of brass = ________________ J/kg °C. [3 marks]

Work and power

1. Synoptic

a When a driver of a car applies the brakes, a force of 5200 N brings the car to a stop over a distance of 50 m. Use the equation $W = F s$ to calculate the work done by the braking force.

Work done = ________________ J [1 mark]

Synoptic

The quantity **work**, W, is defined in Topic 5 of the specification. It is useful to include the equation for work done by a force F moving an object through a distance s in this section, because of the connection between work and energy transfer:

$W = F s$

b Describe the changes to the energy stores as the work done by the braking force brings the car to a stop.

____

____ [2 marks]

Remember

You need to learn these two equations for power as they may not be given to you in the exam:

power = $\frac{\text{energy transferred}}{\text{time}}$,

power = $\frac{\text{work done}}{\text{time}}$

Worked Example

Calculate the useful power output of a person lifting a 10 kg box from the ground onto a 2.0 m high shelf in 1.2 s.

Gravitational field strength = 9.8 N/kg.

Energy transferred to the box = *mgh*

$= 10 \times 9.8 \times 2.0 = 196$

$\text{power} = \frac{\text{energy transferred}}{\text{time}}$

$= \frac{196}{1.2} = 163\text{ W}$

Marks gained: [4 marks]

2. Maths Skills

An electric motor driven crane lifts a 600 kg load through a height of 25 m in 49 s. Calculate the motor's useful output power.

Gravitational field strength = 9.8 N/kg.

Useful output power = ____________________ W [4 marks]

Conservation of energy

1. **a** Describe what is meant by **energy dissipation**.

____

____ [1 mark]

b Describe how energy is dissipated from a moving bicycle.

___

___ [1 mark]

2. Maths Skills

a A cyclist is cycling on a long flat road at a steady speed. The combined effects of air resistance and friction dissipate energy to the surroundings at a rate of 150 W. The cyclist's journey takes 20 minutes. Calculate the total amount of work done by the cyclist in maintaining her steady speed for 20-minute journey.

Work done = ___ J [2 marks]

Maths Skills

b The human body can transfer only 25% of the energy obtained from food to useful work. Calculate the energy from food that the cyclist must consume at the end of the journey to increase her body's store of chemical energy back to its original level.

Energy from food = ___ J [1 mark]

Ways of reducing unwanted energy transfers

1. Required practical

A group of students use the apparatus in **Figure 3** to compare the effectiveness of different materials as thermal insulators.

The apparatus consists of a small beaker positioned inside a large beaker. The gap between the two beakers is filled with insulating material. Hot water is put in the small beaker and its temperature is recorded as time passes.

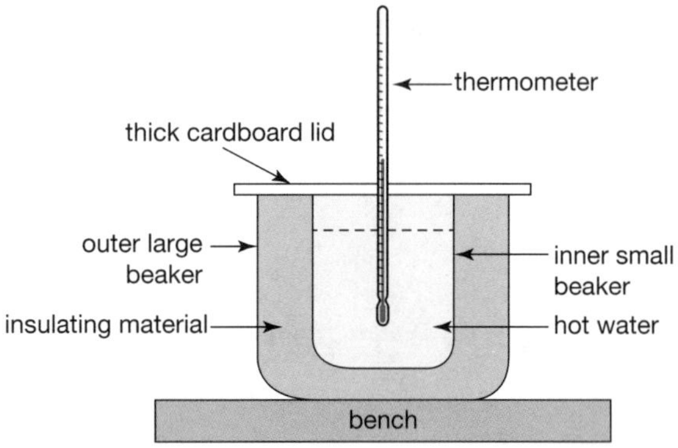

thermometer
thick cardboard lid
outer large beaker
inner small beaker
insulating material
hot water
bench

Figure 3

a Suggest three materials that would be suitable insulators for testing.

___ [1 mark]

b Identify the dependent variable and the independent variable.

___

___ [2 marks]

c Identify **three** control variables.

____

____ [3 marks]

d Describe how the students can keep each of the control variables constant. Refer to any additional apparatus that may be required.

____

____

____

____ [3 marks]

e Describe how the data produced in the experiment should be analysed in order to compare the effectiveness of the insulators tested.

____

____

____ [2 marks]

Efficiency

1. Maths Skills

A car with a petrol engine has an efficiency of 15%. When full, the car's tank is a store of 800 MJ of chemical energy. Calculate the car's useful output energy transfer in consuming a full tank of petrol.

Remember

You need to know the equations for efficiency as they may not be given to you in the exam.

Useful output energy transfer = ______________ [2 marks]

2. A student uses the apparatus in **Figure 4** to measure the efficiency of an electric motor. A mass of 0.50 kg is raised by an electric motor through a height of 0.80 m. The energy supplied to the motor is measured by a joulemeter and is 10.3 J.

Gravitational field strength = 9.8 N/kg.

Maths

Efficiency values are often given as percentages. It is useful to remember that percentages are fractions of 100. So 40% is the fraction $\frac{40}{100}$.

Maths Skills

a Calculate the increase in the gravitational potential energy of the mass.

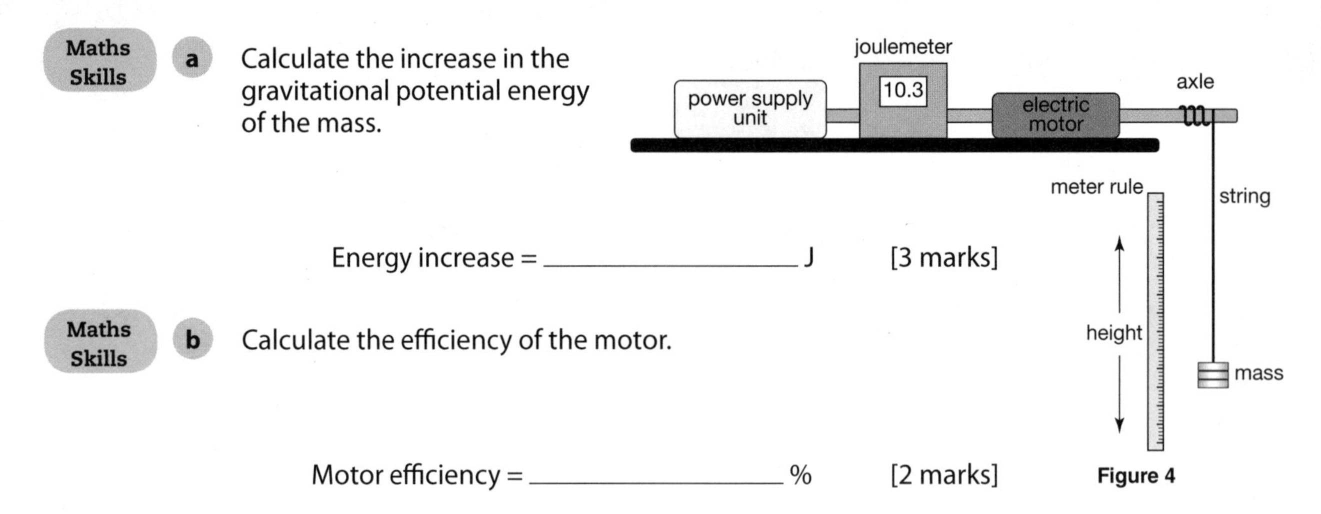

Figure 4

Energy increase = _______________ J [3 marks]

Maths Skills

b Calculate the efficiency of the motor.

Motor efficiency = _______________ % [2 marks]

National and global energy resources

1. The graph in **Figure 5** shows how the demand for electrical power varies during a typical winter's day and a typical summer's day in Great Britain.

Command word

'Suggest' means applying your knowledge and understanding to a new situation.

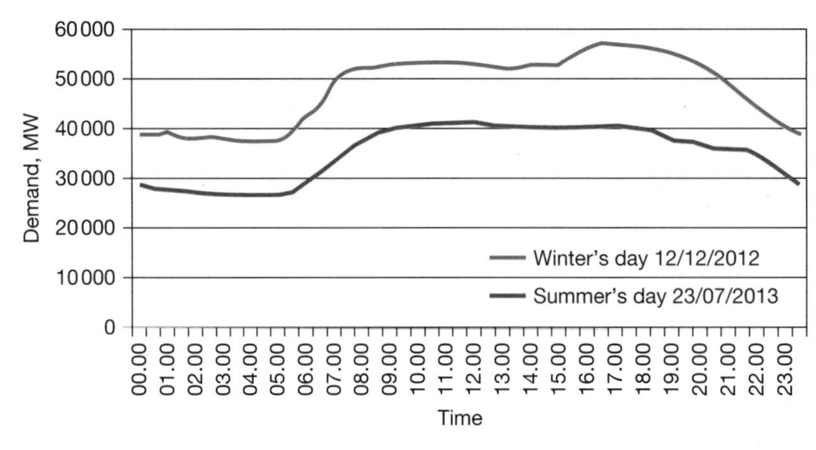

Figure 5

a Suggest why the demand for electricity between 11 p.m. and 7 a.m. is steady and at its lowest both in the summer and the winter.

____

_______________ [2 marks]

b Suggest why the demand for electricity increases between 7 a.m. and 9 a.m. both in the summer and the winter.

____

_______________ [2 marks]

Circuit diagrams

1. Answer parts (a) and (b) by selecting circuit W, X, Y or Z from **Figure 1**.

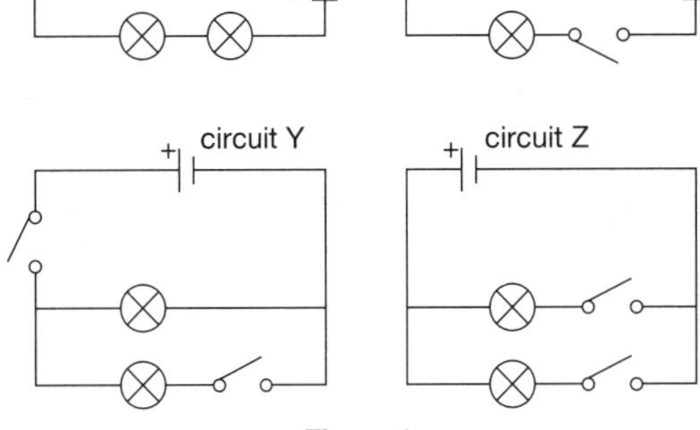

Figure 1

a Which circuit contains two lamps that can be switched on and off independently?

Circuit Z [1 mark]

b In which circuit can the brightness of the lamps be varied?

Circuit X [1 mark]

2. A student is asked to build the circuit shown in **Figure 2**.

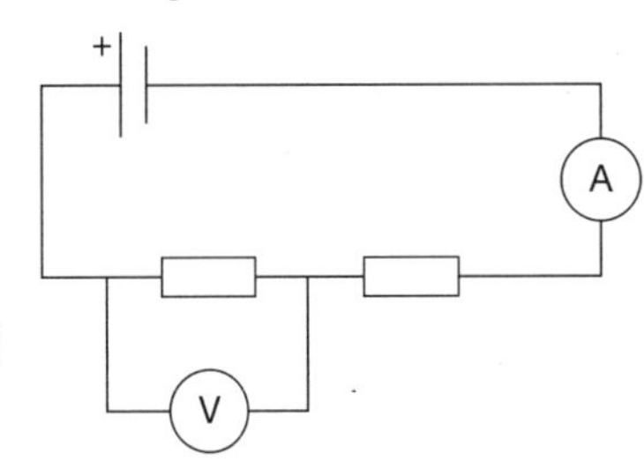

Figure 2

What is the lowest number of connecting leads that the student would need to build the circuit?

Number of leads = 7 or 6 [1 mark]

Electrical charge and current

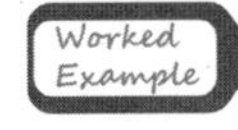

3.0 C of charge flows through a lamp in 150 s. Calculate the current through the lamp. Give your answer in milliamps (mA).

$Q = It$ can be rearravnged to give $I = \frac{Q}{t}$.

Substituting data gives $I = \frac{3.0}{150}$

$= 0.020$ A, which is 20 mA.

Marks gained: [3 marks]

Remember
You need to learn the equation relating charge flow to current and time as it may not be given to you in the exam.

1. A student builds the electric circuit shown in **Figure 3**.

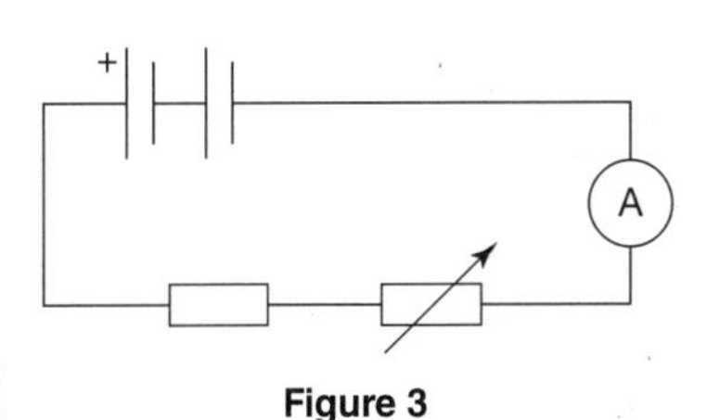

Figure 3

a What is the source of potential difference in the circuit?

the battery [1 mark]

Maths Skills

b The ammeter reads 75 mA. Calculate the charge that flows through the ammeter in 10 s. Use the correct equation from the Physics Equations Sheet.

Charge = 0.150 C [2 marks]

Maths

Electric current values are sometimes given in microamps (μA). 'micro' is known as a prefix and is equal to $\frac{1}{1000000}$. So 1 μA can be written as 1×10^{-6} A.

2. Ammeter A1 in **Figure 4** reads 55 μA.

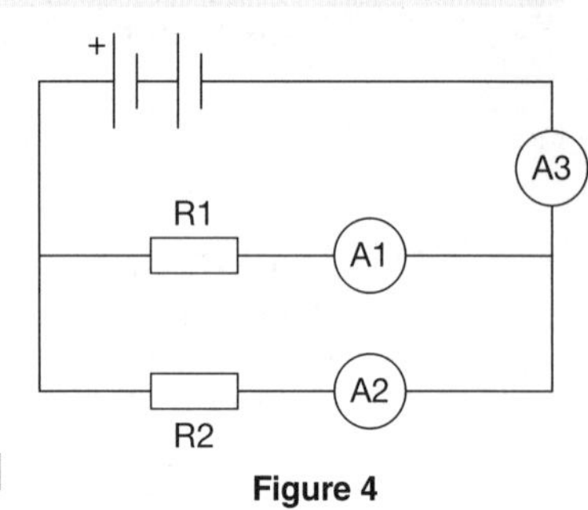

Figure 4

Maths Skills

a Calculate the charge that flows through resistor R1 in 10 minutes. Use the correct equation from the Physics Equations Sheet. 55×10^{-6}

Charge = ______________ C [2 marks]

b A charge of 0.048 C flows through resistor R2 in 10 minutes. Calculate the reading on ammeter A2. Give your answer in microamps (μA).

Ammeter reading = ______________ μA [3 marks]

c Calculate the reading on ammeter A3. Give your answer in microamps (μA).

Ammeter reading = ______________ μA [2 marks]

Electrical resistance

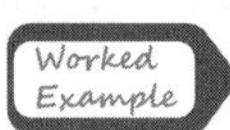

The potential difference across a 180 kΩ resistor is 3.6 V. Calculate the current flowing through the resistor. Give your answer in microamps (μA).

$V = IR$ can be rearranged to

$$I = \frac{V}{R}$$

$$= \frac{3.6}{180 \times 10^{3}}$$

$$= 2.0 \times 10^{-5} \text{ A} = 20 \text{ μA}$$

Remember

You need to know the equation linking potential difference, current and resistance as it as it may not be given to you in the exam.

Marks gained: [3 marks]

1. **Maths Skills**

The potential difference across a 220 kΩ resistor is 3.3 V. Calculate the current flowing through the resistor. Use the correct equation from the Physics Equations Sheet. Give your answer in microamps (μA).

$I = \frac{V}{R}$ $\frac{3.3}{220 \times 10^3} = 1.5 \times 10^{-5} = 15$

Current = 15 μA [3 marks]

2. **a** A student has to find the resistance of two resistors connected in parallel.

Complete the circuit diagram in **Figure 5** to show the required circuit. [3 marks]

Figure 5

b Describe how the student should determine the value of the combined resistance.

____

______________________________ [1 mark]

3. A 3.00 V battery is connected to the ends of a length of constantan wire of resistance 12.1 Ω. Calculate the charge that flows through the wire in 60 s. Write down any equations you use. Give your answer to three significant figures.

Charge = ______________ C [5 marks]

Problem solving

In a 5 or 6 mark calculation question there is usually more than one problem to solve to get the information needed to calculate the final answer. Always show your working so that you get some marks even if your final answer is wrong.

4. **Required practical**

A student is asked to find out how the resistance of a piece of constantan wire depends on its length. She is given the apparatus shown in **Figure 6**. The full length of the wire is about 120 cm. The crocodile clips, X and Y, can be moved to change the length of wire connected in the circuit.

a Identify the variables.

Independent variable: ______________

Dependent variable: ______________

Control variable: ______________ [3 marks]

+
A
X
wire
Y
V

Figure 6

b Describe a plan that the student should follow to generate sets of data for the resistance of various lengths of the piece of wire connected between X and Y. Refer to any additional apparatus required. Include advice on how to improve the accuracy of the results and also how to prevent the wire from overheating.

Literacy

An experiment plan should be written so that the steps involved are in a logical sequence that another person could follow and get valid results.

[6 marks]

c Identify the variables of the graph that the student should plot to determine the relationship between resistance and length of wire.

Variable on *x*-axis: ____________________

Variable on *y*-axis: ____________________ [2 marks]

d The student's data showed that the resistance of a wire is directly proportional to its length. Describe the main features of the graph that would allow this conclusion to be drawn.

[2 marks]

Resistors and *I*–*V* characteristics

1. The circuit in **Figure 7** contains a light-dependent resistor (LDR). The light incident on the LDR is measured with a light meter which records the brightness of the light in units of lux. The greater the brightness the larger the lux reading on the light meter. When the switch is closed, the ammeter and voltmeter readings are recorded both with and without a torch shining on the LDR.

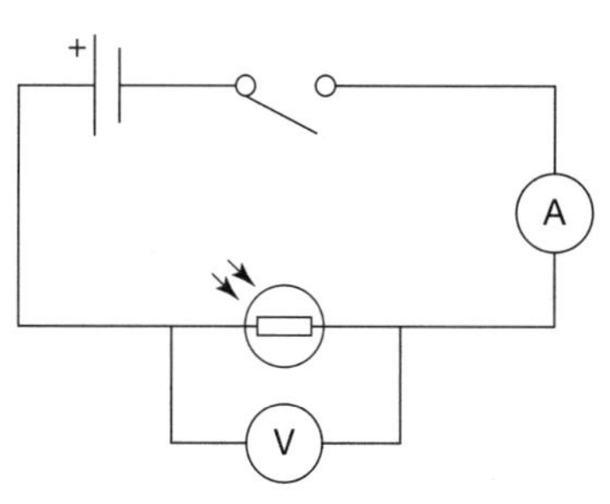

Figure 7

Maths Skills

a The readings are shown in the table.

Light meter reading (lux)	Current (mA)	Potential difference (V)	LDR resistance (Ω)
1800	75.0	6.0	
150	2.4	6.0	

Complete the table by calculating the LDR's resistance values. Use the correct equation from the Physics Equations Sheet. [4 marks]

b What can be concluded from the measurements about how the resistance of an LDR can be changed?

____

______________________________ [2 marks]

2. A student measures the resistance of a thermistor using an ohmmeter. He puts the thermistor in a beaker of hot water and measures its resistance as the water cools. He plots a graph of resistance of the thermistor versus time (**Figure 8**).

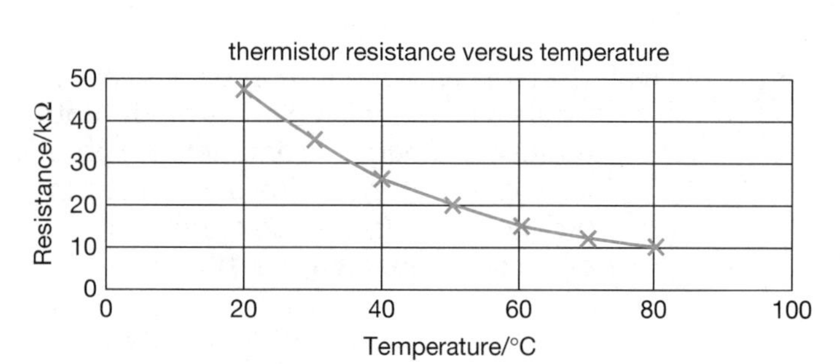

Figure 8

Maths Skills

a Use **Figure 8** to determine the resistance of the thermistor at 25°C and at 55°C. Show on **Figure 8** how you work out your answer.

Resistance at 25°C = ______________ Ω

Resistance at 55°C = ______________ Ω [2 marks]

b What can be concluded from the graph about how the resistance of a thermistor can be changed?

____

______________________________ [2 marks]

3. **a** State what is meant by an ohmic conductor.

____

______________________________ [2 marks]

Maths Skills

b A fixed resistor is classed as an ohmic conductor. When the potential difference across a particular fixed resistor is 6.0 V, the current through the resistor is 0.80 A. Calculate the current that would flow through the resistor if the potential difference is increased to 9.0 V. Use the correct equation from the Physics Equations Sheet.

Current = ______________ A [2 marks]

4. **a** A student is asked to obtain current readings at different voltages for two different electrical components.

Required practical

In **Figure 9** the component is represented by the box labelled X. Complete the circuit diagram in **Figure 9** to show the required circuit. Include a switch in your circuit diagram.

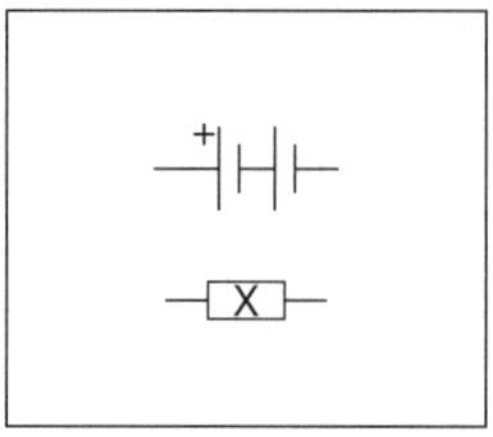

Figure 9

[4 marks]

b The two components that data is collected for are a filament lamp and a diode. Both positive and negative data sets are obtained for each component. The data is used to produce current–voltage graphs for each component (**Figure 10**).

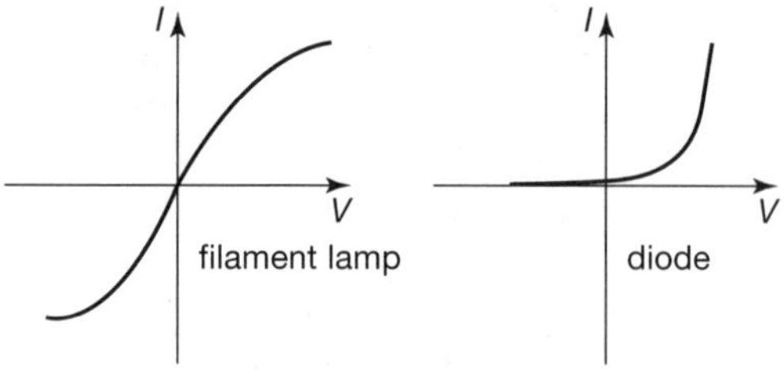

Figure 10

Explain how resistance changes can be deduced from a current–voltage graph.
In your answer, use the graphs to compare how the resistances of the lamp and the diode vary as the potential difference is changed.

Command word

When asked to 'compare' you should describe the similarities and/or differences between things.

[6 marks]

Series and parallel circuits

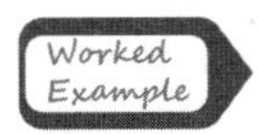

Calculate the reading on the ammeter in **Figure 11** when the switch is open and when the switch is closed.

When the switch is open the circuit contains two 4.0 Ω resistors in series. Their total resistance = 8.0 Ω.

Ammeter reading = $I = \frac{V}{R}$

$= \frac{6.0}{4.0 + 4.0}$

$= 0.75$ A

When the switch is closed it acts as a short circuit. The 4.0 Ω resistor that the switch is connected across is effectively removed.

The ammeter reading = $I = \frac{V}{R}$

$= \frac{6.0}{4.0} = 1.5$ A

Marks gained: [5 marks]

Remember

The total resistance of components connected in series is the sum of their resistances.

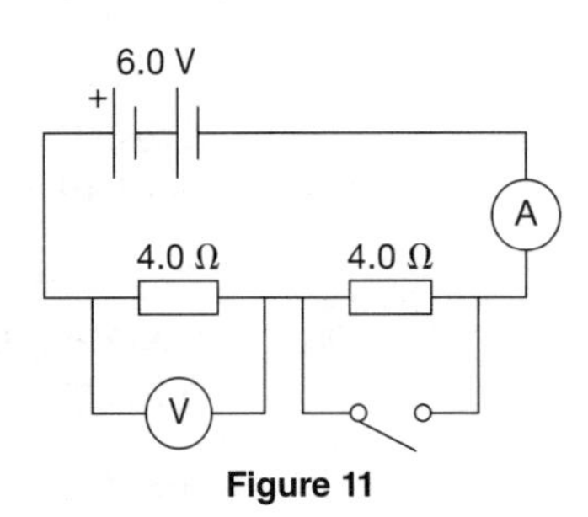

Figure 11

1. Calculate the readings on the voltmeter in **Figure 11** when the switch is open and when the switch is closed. Use the correct equation from the Physics Equations Sheet.

Voltmeter reading (switch open) = ________________ V

Voltmeter reading (switch closed) = ________________ V [4 marks]

2. **a** The three resistors in **Figure 12** are identical and each has a resistance of 2.0 kΩ.

Remember

When components are connected in parallel, the potential difference across each component has the same value.

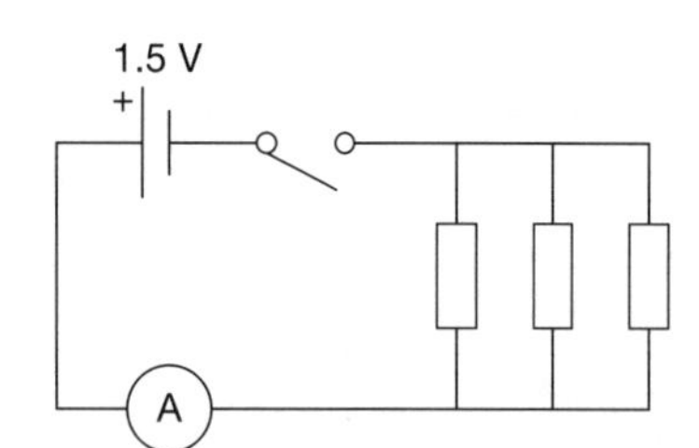

Figure 12

Calculate the current flowing through one of the 2.0 kΩ resistors when the switch is closed. Use the correct equation from the Physics Equations Sheet. Give your answer in milliamps (mA).

Current = ________________ mA [2 marks]

b Calculate the reading on the ammeter when the switch is closed.

Give your answer in milliamps (mA).

Ammeter reading = ________________ mA [2 marks]

3. The fixed resistor in **Figure 13** has a resistance of 18 kΩ. At 60 °C, the thermistor has a resistance of 12 kΩ, and at 20 °C it has a resistance of 42 kΩ.

a Calculate the reading on the voltmeter when the thermistor is at a temperature of 60 °C.

Write down any equations you use.

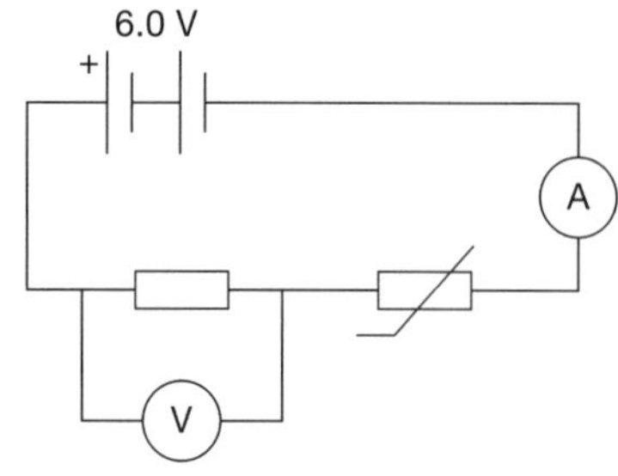

Figure 13

Voltmeter reading = ________________ V [4 marks]

b Calculate the reading on the voltmeter when the thermistor is at a temperature of 20 °C.

Voltmeter reading = ________________ V [4 marks]

Mains electricity

1. Give the potential difference and frequency of the mains electricity supply in the UK.

Potential difference = ______________ V   Frequency = ______________ Hz

[2 marks]

2. An electric toaster contains heating elements made up of nichrome wire. The wires get hot when they carry an electric current. Sometimes a fault can occur in which some of the nichrome wire makes contact with the metal casing of the toaster. Explain how the earth wire connecting the toaster to the plug prevents someone using the toaster receiving an electric shock.

Command word

Explain means you have to write what happens and **why** it happens.

____

____

____

____

______________________________ [4 marks]

Energy changes in circuits

1. State the two factors that determine the energy transferred by a household electrical appliance.

____

______________________________ [2 marks]

2. A 1.2 kW hairdryer transfers energy from the mains electricity supply to other energy stores.

Maths Skills

Calculate the energy transferred from the mains in 10 minutes of operation. Use the correct equation from the Physics Equations Sheet.

Remember

You need to learn the equation that relates energy transferred to power and time as you may not be given the equation in the exam.

Energy = ______________ [2 marks]

3. **a** An LCD television operates at a power of 1.5 W when in standby mode. The television is left in standby mode for an average of 6 hours a day. Calculate the energy transferred from the mains electricity supply in 1 year by leaving the television in standby mode. Use the correct equation from the Physics Equations Sheet. Give your answer in standard form to two significant figures.

Energy = ____________________ J [3 marks]

b In 1 hour, a 1 kW device transfers 3.6×10^6 J from the mains electricity supply (known as one kilowatt-hour). The cost of using this amount of energy is about 15 pence. Calculate the cost of the energy calculated in part (a).

Cost = ____________________ [2 marks]

Worked Example

A 1.5 kW dishwasher is switched on for 1 hour. Calculate how long a 90 W LED television would have to be switched on to transfer the same amount of energy from the mains electricity supply.

Energy transferred by the dishwasher in 1 hour = power × time

$= 1500 \times 3600$

$= 5.4 \times 10^6$ J

For the television, time $= \dfrac{\text{energy transferred}}{\text{power}}$

$= \dfrac{5.4 \times 10^6}{90}$

$= 6.0 \times 10^4$ s

$= \dfrac{6.0 \times 10^4}{60} = 1000$ minutes

Marks gained: [4 marks]

Maths

You could also work out the answer to this question using ratios. Whichever method you use, always show your working. You may still get some marks if your final answer is wrong.

4. **Maths Skills**

A personal computer with a power rating of 100 W is switched on for 20 minutes. Calculate how long a 12 W energy-efficient light bulb would have to be switched on to transfer the same amount of energy from the mains electricity supply. Use the correct equation from the Physics Equations Sheet. Give your answer in minutes.

Time = ____________________ minutes [4 marks]

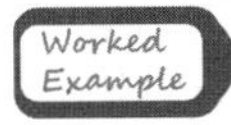

An LED torch operates at a power of 0.50 W when connected to a 3.0 V battery. Calculate the amount of charge that flows through the torch when it is switched on for 10 minutes.

Remember

You need to know the equation that links energy transferred to charge flow and potential difference as you may not be given this equation in the exam.

Energy transferred, $E = Pt$

$= 0.50 \times 10 \times 60$

$= 300$ J

Energy transferred, $E = QV$, can be rearranged to give $Q = \frac{E}{V}$

$= \frac{300}{3.0} = 100$ C

Marks gained: [3 marks]

5. Maths Skills

a An electric oven takes a current from the 230 V mains electricity supply. The oven has a power of 2.5 kW. Calculate the energy transferred to the oven in 50 minutes.

Use the correct equation from the Physics Equations Sheet. Give your answer in standard form.

Energy transferred = ______________ J [3 marks]

b Calculate the total charge that flows through the oven's heating element in this time.

Give the answer in standard form to two significant figures.

Charge = ______________ C [3 marks]

Electrical power

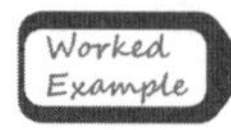

Calculate the electric current drawn from the 230 V mains electricity supply by an 800 W microwave oven.

The equation $P = VI$ can be rearranged to give $I = \frac{P}{V}$

$$= \frac{800}{230} = 3.5 \text{ A}$$

Marks gained: [2 marks]

Remember

You need to learn the equation that relates electrical power to potential difference and electric current as you may not be given the equation in the exam.

1. Maths Skills

Calculate the electric current drawn from the 230 V mains electricity supply by a 1.1 kW vacuum cleaner. Use the correct equation from the Physics Equations Sheet. Give your answer to two significant figures.

Current = ______________ A [3 marks]

2. Maths Skills

The heating element of a 2.3 kW electric kettle takes a current from the 230 V mains electricity supply. Calculate the resistance of the heating element.

Resistance = ______________ Ω [4 marks]

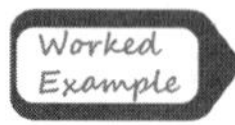

A resistor connected to a battery transfers energy from the chemical store of the battery to thermal energy. Calculate the power dissipated by a 1.0 kΩ resistor if it is carrying a current of 4.0 mA.

Power, $P = I^2R$

$$= (4.0 \times 10^{-3})^2 \times 1000$$

$$= 0.016 \text{ W}$$

Marks gained: [2 marks]

Remember

You need to learn the equation that relates power to current and resistance as you may not be given the equation in the exam.

3. Maths Skills

A 50 m long extension cable has a resistance of 1.0 Ω. When the cable connects a lawnmower to the mains electricity supply it carries an electric current of 3.0 A. Calculate the power dissipated by the extension cable.

Power = ______________ W [2 marks]

The National Grid

1. **a** The National Grid consists of many thousands of kilometres of overhead power line.

Each power line consists of a bundle of very thick cables. The current flowing in a power line transfers some energy as thermal energy to the surroundings. Compare the power dissipated as thermal energy by a power line carrying a current of 100 A with an identical power line carrying a current of 1000 A. Assume that a 1 km length of power line has a resistance of about 0.1 Ω. Write down any equations you use.

____

____

____

____

____

____

____

____ [6 marks]

b State how electrical power should be transmitted to minimise the energy dissipated as thermal energy to the surroundings.

____

____ [1 mark]

2. **a** In terms of potential difference, describe the difference between the action of a step-up transformer and a step-down transformer.

____

____ [1 mark]

b **Figure 14** is a simplified representation of the National Grid. Identify A and B as either step-up or step-down transformers.

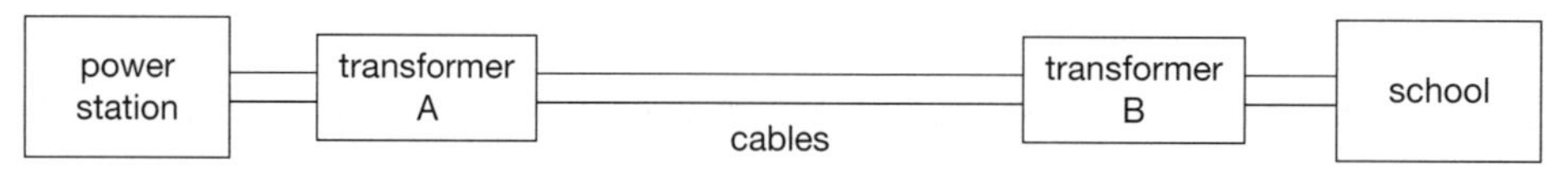

Figure 14

Transformer A ____________________

Transformer B ____________________ [2 marks]

Synoptic
There are more questions on transformers in Chapter 7.

Static electricity

1. **a** A student finds that when he rubs an acetate rod with a cloth, the rod becomes positively charged. Explain how the acetate rod becomes positively charged.

____

____ [2 marks]

b Describe the effect this rubbing has on the cloth.

____ [1 mark]

c The student suspends the charged acetate rod so that it is free to rotate (**Figure 15**). The student is given a second acetate rod, a polythene rod and suitable cloths with which to charge both rods. Note that the **polythene** rod becomes **negatively** charged when rubbed with a cloth.

Describe how the student could use the apparatus to demonstrate that there are two types of charge.

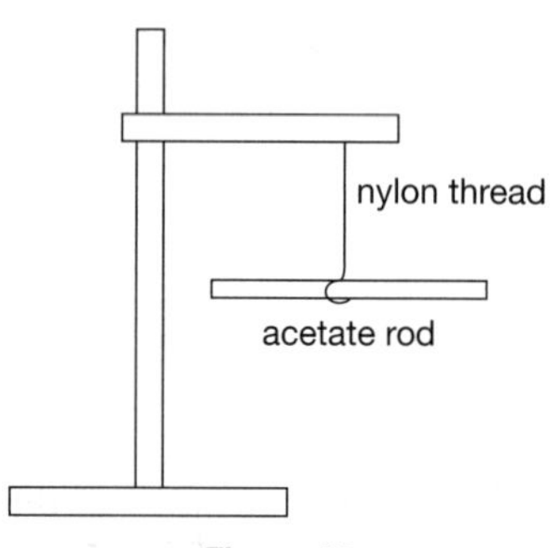

Figure 15

____

____

____

____

____ [3 marks]

Electric fields

1. **a** **Figure 16** shows the electric field around a charged insulated metal sphere.

State what type of charge is on the sphere.

Charge: ______________ [1 mark]

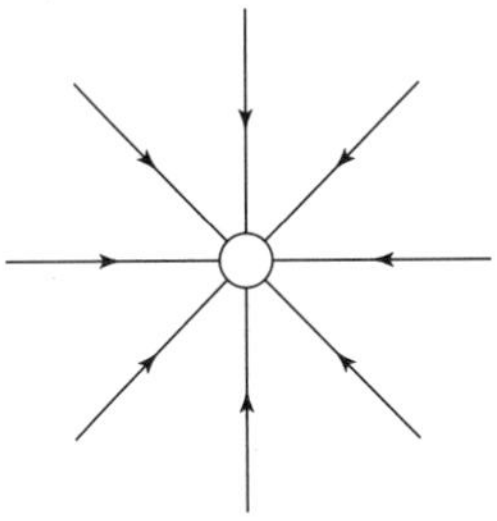

Figure 16

b Describe the effect that the electric field would have on a negatively charged speck of dust in the air close to the sphere.

____

______________ [1 mark]

2. A light, charged ball is moved towards the charged dome of a Van de Graaff generator (**Figure 17**).

a State whether the force between the ball and the dome is attractive or repulsive.

Force: ______________ [1 mark]

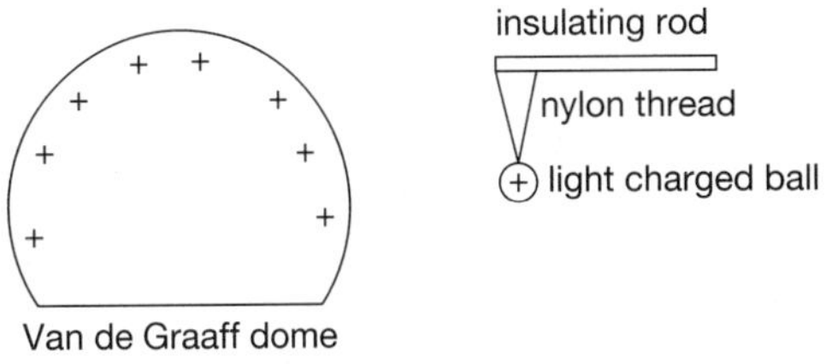

insulating rod

nylon thread

light charged ball

Van de Graaff dome

Figure 17

b Describe how the force on the ball changes as it is moved closer to the dome.

______________ [1 mark]

Density

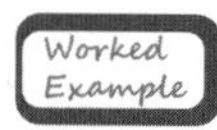

One of the densest materials seen in everyday life is gold. Gold has a density of 19 300 kg/m^3. Calculate the mass of a cube of pure gold if the cube has sides of 1.00 cm.

Volume = $(0.01 \text{ m})^3$

$= 1.0 \times 10^{-6} \text{m}^3$

Mass, $m = \rho V$

$= 19300 \frac{\text{kg}}{\text{m}^3} \times 1.0 \times 10^{-6} \text{m}^3$

$= 0.0193 \text{ kg}$

Marks gained: [4 marks]

Maths
You need to know the equation relating density to mass and volume as you may not be given the equation in the question.

1. **Required practical**

a A student measures the dimensions of a small rectangular block of metal using a 30 cm rule graduated in mm. He also measures the mass of the block using a digital balance. The student's measurements are shown in **Table 1**. **Table 2** gives the densities of some metals. Show that the block is probably made of steel.

Length (cm)	Width (cm)	Height (cm)	Mass (g)
8.0	5.0	2.0	624

Table 1

Metal	Zinc	Tin	Steel	Nickel
Density (kg/m^3)	7140	7280	7850	8910

Table 2

[4 marks]

b Give **two** reasons why the block cannot be conclusively identified as steel from the student's measurements.

____

____

____

____ [2 marks]

c Give a suitable estimate of the uncertainty on the measurements of the block's dimensions made with the 30 cm rule. Give a unit with your answer.

Uncertainty = ________________ unit: ________________ [1 mark]

2. **Required practical**

A student is asked to determine the density of a pebble using the apparatus in **Figure 1**, along with an electronic balance. The measuring cylinder has a capacity of 50 cm^3 and has graduations of 0.5 cm^3.

Figure 1

a Describe the steps that the student should follow to determine the density of the pebble. You can assume that a piece of cotton is available along with an additional glass beaker.

Literary

Answers to 4- and 6-mark extended written questions should be written in sentences and in an order that makes sense.

____

____

____

____

____

____

____

____

____ [4 marks]

Maths Skills

b The measurements for the mass and volume of the pebble are 84.2 g and 38.5 cm^3. Calculate the density of the pebble. Give your answer to three significant figures.

Density = ______________ g/cm^3 [2 marks]

c Another student suggests that instead of lowering the pebble into the displacement can and collecting the displaced water in the measuring cylinder it would be simpler to lower the pebble into a measuring cylinder. It would then be possible to measure the pebble's volume from the difference between the initial and final water levels. A measuring cylinder wide enough to fit the pebble would typically have a capacity of 250 m^3 with graduations of 5 cm^3. Explain what effect this alternative method would have on the uncertainty in the calculated value for the pebble's density.

____

____

______________________________ [3 marks]

Changes of state

1. Name each of the following processes.

A liquid, at a temperature below its boiling point, changing to a gas:

______________________________ [1 mark]

A solid changing directly into a gas:

______________________________ [1 mark]

Command word

Compare means to describe the similarities and the difference between things.

2. Compare the motion and the arrangement of the particles of a substance in the solid state with the same substance when it has melted and is in the liquid state.

____

____

____

______________________________ [3 marks]

3. A solid substance is heated from room temperature to 100 °C. During the heating the substance changes state. It is then allowed to cool and its temperature is measured as time passes (**Figure 2**).

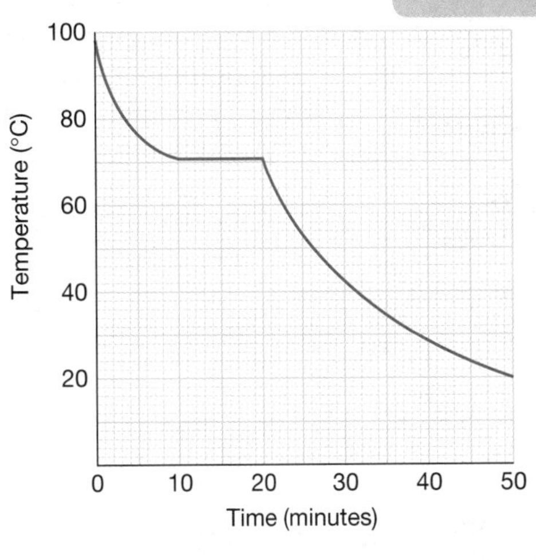

Figure 2

a Give the time interval during which a change of state takes place.

Time interval: ______________________ [1 mark]

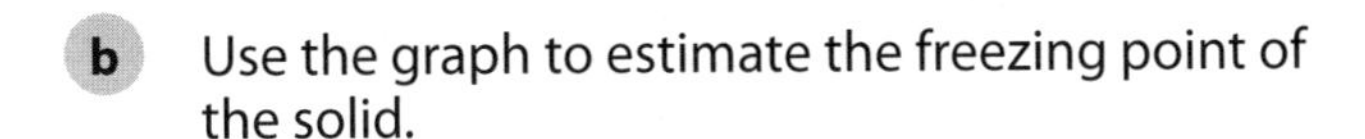

b Use the graph to estimate the freezing point of the solid.

Melting point = ______________________ °C [1 mark]

Internal energy and specific latent heat

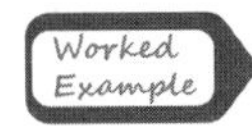

100 g of ice, at −20 °C, is removed from a freezer. Calculate the energy transferred from the room to the ice to change it to water at 0 °C. Specific heat capacity of ice = 2100 J/kg °C. Specific latent heat of fusion of water = 336 kJ/kg. Write down the equations you use. Give the answer in kJ to two significant figures.

To raise the temperature of the ice to 0 °C:

$$\Delta E = m c \Delta\theta$$

$$= 0.1 \times 2100 \times 20$$

$$= 4200 \text{ J}$$

To change the state of the ice:

$$E = m L$$

$$= 0.1 \times 336\,000$$

$$= 33\,600 \text{ J}$$

Total energy = 33 600 + 4200

= 37 800 J = 38 kJ

Marks gained: [6 marks]

Maths

The equations: $\Delta E = m\,c\,\Delta\theta$ and $E = m\,L$ are on the Physics Equation Sheet. You need to be able to select and apply the first equation for the energy change involved in raising and lowering the temperature of a system and the second for the energy change involved in changes of state.

1. Synoptic Maths Skills

a In a steel-making furnace, 1000 kg of cast iron is heated from 20 °C to its melting point of 1200 °C. Calculate the energy transferred to the iron to raise its temperature by this amount. Use the correct equation from the Physics Equations Sheet. Give your answer in MJ.

Specific heat capacity of solid iron = 500 J/kg °C.

Energy transferred = ____________________ MJ [3 marks]

Maths Skills

b Calculate the total energy transferred to the cast iron to change it from solid at 1200 °C to liquid at 1200 °C. Use the correct equation from the Physics Equations Sheet. Give your answer in MJ.

Specific latent heat of fusion of cast iron = 10 kJ/kg.

Energy transferred = ____________________ MJ [3 marks]

Maths Skills

c Calculate the total energy required to change the solid cast iron at 20 °C to molten iron at 1200 °C. Give your answer in MJ.

Energy required = ____________________ MJ [2 marks]

2. a A glass beaker contains tap water at room temperature. The water is described as having internal energy. Describe the **two** components of the water's internal energy.

____

____ [1 mark]

b The beaker of water is now heated with a Bunsen burner. State which of the **two** components of internal energy increases as the water temperature rises.

____

____ [1 mark]

3. A heating curve for a substance is shown in **Figure 3**. The data has been obtained by supplying energy to the substance at a constant rate in apparatus insulated from its surroundings.

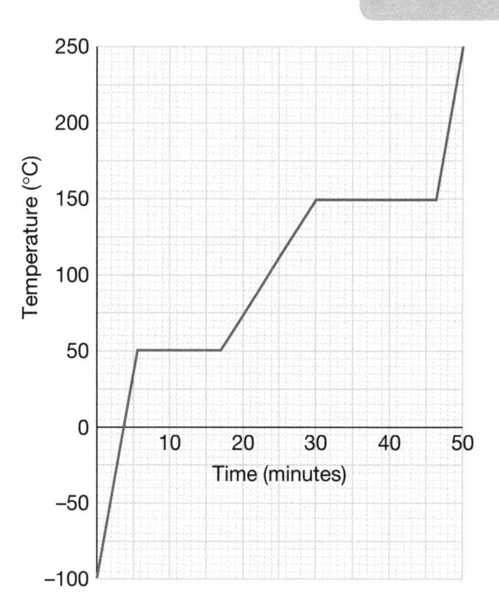

Figure 3

a Determine the substance's melting point and boiling point.

Melting point = ____________________ °C

Boiling point = ____________________ °C

[2 marks]

b Using information from the graph, state which is greater – the substance's specific heat capacity when in the solid state or in the liquid state. Give a reason for your answer.

____

____

____ [2 marks]

Maths

The steeper the gradient of a temperature versus time graph, the quicker the temperature is changing.

4. A student uses the apparatus in **Figure 4** to determine the specific latent heat of vaporisation of water.

The water is heated by the immersion heater. When the water is boiling briskly, the joulemeter and the balance readings are recorded. After 10 minutes the readings are taken again. The readings taken by the student are in the table. Use the data to determine a value for the specific latent heat of vaporisation of water. Give your answer in J/kg in standard form. Write down the equations you use.

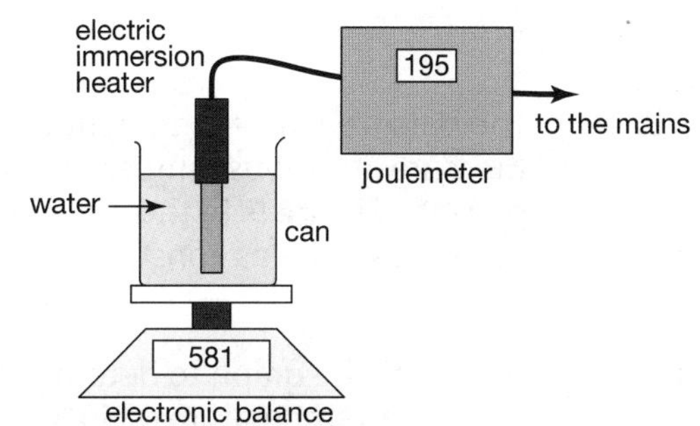

Figure 4

First balance reading (g)	Second balance reading (g)	First joulemeter reading (kJ)	Second joulemeter reading (kJ)
581	526	195	327

Specific latent heat = ____________________ J/kg [5 marks]

Particle motion in gases

1. The particle model of a gas describes gas molecules as moving with a range of speeds in different directions.

a Explain, in terms of the motion of the molecules, how a gas in a fixed container exerts pressure on the inside walls of the container.

____

____

______________________________ [2 marks]

b Give **two** reasons why increasing the speed of the molecules would cause an increase in the pressure exerted by the gas.

____

____

______________________________ [2 marks]

2. The data from an experiment to measure gas pressure at various temperatures is shown on the graph in **Figure 5**. In the experiment, the volume of the gas remains constant.

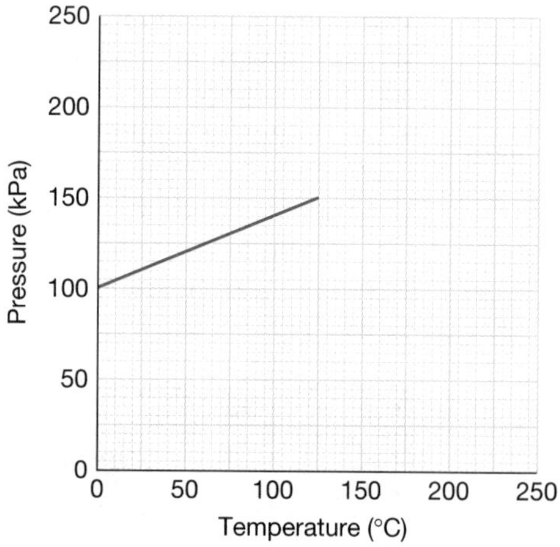

a Use the graph to determine the gas pressure at a temperature of 50 °C.

Pressure = ______________ kPa [1 mark]

b Use the graph to predict the gas pressure at a temperature of 250 °C. Show on the graph how you obtained the value. State any assumption that you make.

Pressure (kPa): 0, 50, 100, 150, 200, 250
Temperature (°C): 0, 50, 100, 150, 200, 250

Figure 5

Assumption: ______________________________

____

Pressure = ______________ kPa [3 marks]

Increasing the pressure of gas

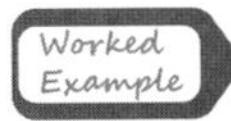

A quantity of gas has a volume of 0.20 m^3 at a pressure of 80 000 Pa. The gas is slowly compressed, at a constant temperature, until its volume is 0.16 m^3. Calculate the new gas pressure.

pV = constant

= 80 000 × 0.20 = 16 000

The new pressure: $p_{new} = \frac{\text{constant}}{V_{new}}$

$= \frac{16000}{0.16}$

= 100 000 Pa

Marks gained: [3 marks]

Maths

The equation linking the pressure and volume of a gas at constant temperature is given on the Physics Equations Sheet. You are expected to select and apply the equation.

1. **Figure 6a** represents a gas-filled cylinder sealed by a moveable piston.

a If the additional weights are removed, the piston moves upwards slowly and then comes to a stop (**Figure 6b**). The temperature of the gas is unchanged.

By referring to the gas molecules explain how the pressure of the gas changes when the additional weights are removed.

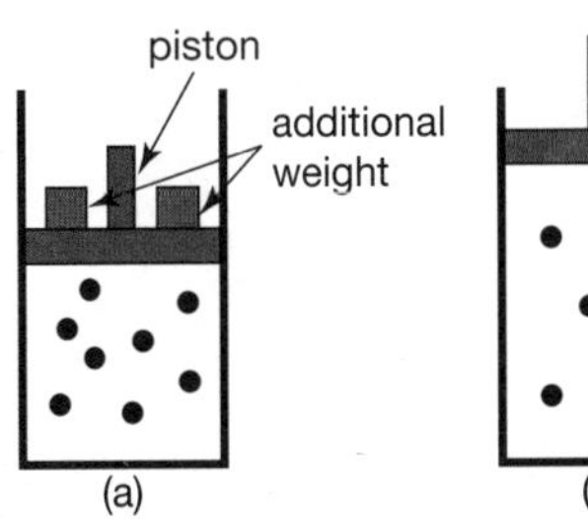

Figure 6

____

____

____

____ [3 marks]

Maths Skills

b The volume of the gas shown in **Figure 6a** is 0.18 m^3 and the gas pressure is 2.5×10^5 Pa. The volume of the gas in **Figure 6b** is 0.25 m^3. Calculate the pressure of the gas shown in **Figure 6b**. Use the correct equation from the Physics Equations Sheet. Give your answer in standard form.

Pressure = ________________ Pa [5 marks]

2. **Figure 7** shows a bicycle pump which can be used to pump air into a bicycle tyre. When a force is applied to push the piston further into the barrel, work is done on the air trapped below the piston. Gas molecules in the trapped air bounce off the moving piston at an increased speed. Explain any change to the temperature, the internal energy and the pressure of the trapped air as a result of the work done in pushing the piston.

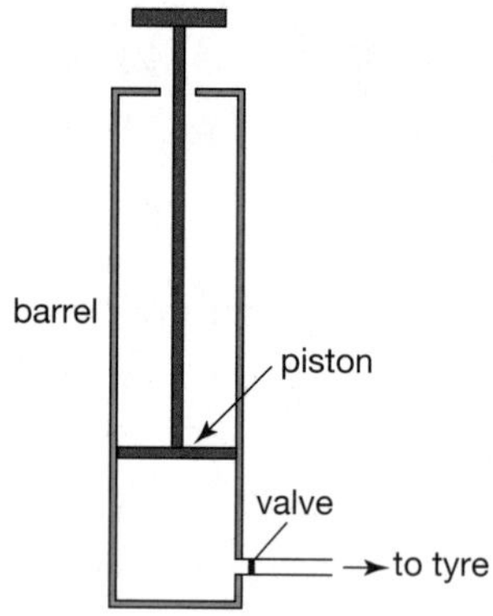

barrel

piston

valve

to tyre

Figure 7

____

____

____

____

____

____

____ [4 marks]

Protons, neutrons and electrons

1. **a** **Figure 1** represents a helium atom. Complete the labels to identify the particles present in a helium atom. [2 marks]

proton

Figure 1

b State which of the particles in the atom are charged and state whether their charge is positive or negative.

____

________________________________ [4 marks]

2. The table gives the mass values for a proton, a neutron and an electron, to two significant figures.

Particle	Proton	Neutron	Electron
Mass (kg)	1.7×10^{-27}	1.7×10^{-27}	9.1×10^{-31}

Maths Skills **a** An atom of carbon has 6 protons, 6 neutrons and 6 electrons. Calculate the mass of the nucleus of the atom. Give your answer to two significant figures in standard form.

Mass = ________________ kg [3 marks]

Maths Skills **b** Calculate how many times more massive a proton is than an electron. Give your answer to the nearest hundred.

Number of times more massive = ________________ [2 marks]

The size of atoms

1. Maths Skills **a** An experiment to determine the radius of an atom was first done around 1890. A value of ~1×10^{-10} m was obtained. In 1911 an estimate for the radius of a nucleus was determined as ~1×10^{-14} m. Calculate how many orders of magnitude greater the radius of an atom is compared with the radius of a nucleus.

Orders of magnitude greater = ________________ [2 marks]

Maths

An order of magnitude is an estimate rounded to the nearest power of 10. When two numbers are the same order of magnitude this means that one is less than 10 times as large as the other.

Maths Skills

b In the 20th century, more accurate values for the radius of an atom and a nucleus were determined. The radius of an oxygen atom is 6×10^{-11} m and the radius of its nucleus is 3×10^{-15} m. Calculate how many times bigger the radius of an oxygen atom is compared with the radius of its nucleus.

Number of times bigger = ____________________ [2 marks]

Elements and isotopes

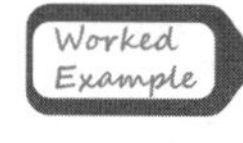

The symbol for the nucleus of one of the isotopes of uranium is $^{238}_{92}U$. Determine the number of protons, neutrons and electrons in one atom.

The atomic number gives the number of protons and also the number of electrons in a neutral atom. The number of neutrons is found by subtracting the atomic number from the mass number.

Protons = 92; electrons = 92; neutrons = 238 − 92 = 146

Marks gained: [3 marks]

1. There are three naturally occurring isotopes of carbon, which can be represented as: $^{12}_{6}C$, $^{13}_{6}C$ and $^{14}_{6}C$.

Compare the three naturally occurring isotopes of carbon in terms of their masses and subatomic particles.

Command word

Compare means describe the similarities and differences between things.

____

____

____

____

____

____________________ [4 marks]

2. **a** One isotope of gold has 79 protons and 118 neutrons in its nucleus. The chemical symbol for gold is shown below. Add the atomic number and mass number to the symbol.

$\_\_\text{Au}$ [2 marks]

b Give the number of electrons in a neutral gold atom.

Number of electrons = ____________. [1 mark]

Maths Skills

c There are 37 isotopes of gold. The lightest is $^{169}_{79}Au$ and the heaviest is $^{205}_{79}Au$. Calculate how many times heavier a nucleus of $^{205}_{79}Au$ is compared with a nucleus of $^{169}_{79}Au$. Give your answer to two significant figures.

Proton mass is the same as neutron mass = 1.7×10^{-27} kg (to two significant figures).

Number of times heavier = ____________ [6 marks]

Electrons and ions

1. The charges on particles in an atom are shown in the table.

Particle	Proton	Neutron	Electron
Charge	$+1.6 \times 10^{-19}$	zero	-1.6×10^{-19}

Maths Skills

a The symbol for the nucleus of a sodium atom is $^{23}_{11}Na$. Calculate the total charge on the nucleus, the total charge of the orbiting electrons and the overall charge on the atom. Give your answers to two significant figures.

Charge on nucleus = ____________ C

Charge on electrons = ____________ C

Overall atom charge = ____________ C [5 marks]

b The symbol for a sodium ion is $^{23}_{11}Na^{+}$. The sodium ion has one fewer orbiting electrons than a sodium atom. Determine the overall charge of the sodium ion.

Charge = ____________ C [2 marks]

2. **Figure 2** represents a lithium ion. There are four neutrons in the nucleus. Give the overall charge on the lithium ion. Explain how you determined your answer.

Charge = ____________ C

Explanation:

____

____

Figure 2

____ [3 marks]

Discovering the structure of the atom

1. In 1909, physicist Ernest Rutherford decided to investigate the plum pudding model of the atom. He did this by firing alpha particles at a very thin sheet of gold.

a Describe the experimental result that could not be explained by the plum pudding model.

____

____________________ [1 mark]

b The nuclear model of the atom replaced the plum pudding model. Compare the distribution of mass and charge in the plum pudding model with the nuclear model.

____

____

____

____________________ [4 marks]

2. Describe the nuclear model of the atom. In your answer you should include:

- a description of the distribution of mass and charge,
- the names and location of the particles,
- a comparison of the mass and charge of the particles,
- an estimate for the radius of an atom.

Do **not** include a diagram.

____

____

____

____

____

____

____

____

____

____ [6 marks]

Radioactive decay

1. **a** A nucleus of protactinium, $^{234}_{91}Pa$, decays by emitting a beta particle and becomes a nucleus of uranium, $^{234}_{92}U$. State the alternative name for a beta particle.

____ [1 mark]

b Describe how the nucleus of $^{234}_{92}U$ is different to a nucleus of $^{234}_{91}Pa$.

____ [2 marks]

2. **a** A nucleus of cobalt, $^{60}_{27}Co$, decays by emitting a beta particle and becomes a nucleus of nickel. Give the atomic number and mass number of the nickel nucleus.

Atomic number = ____________________

Mass number = ____________________ [2 marks]

b The nickel nucleus is created with too much energy. It gets rid of the excess energy by emitting a gamma ray. State the nature of gamma rays.

____ [1 mark]

Comparing alpha, beta and gamma radiation

1. A Geiger–Muller tube (**Figure 3**) is used to detect ionising radiation. Each time an ionising particle enters the Geiger–Muller tube, it ionises some of the gas atoms inside the tube. The ions and free electrons produced are detected and a counter keeps count of the number of particles entering the tube.

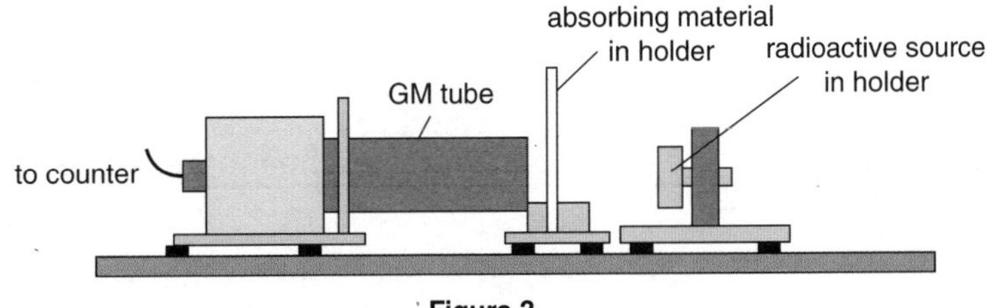

Figure 3

a When no source is present, the counter reads 120 counts in 300 s. These counts are due to background radiation. Calculate the background count-rate in counts per second.

Background count-rate = ________________ per second [2 marks]

b A source of beta particles is placed in the source holder. With no absorber present, 22 300 counts are recorded in 10.0 s. Calculate the count-rate in counts per second.

Count-rate = ________________ per second [1 mark]

c The number of counts is recorded for different thicknesses of aluminium sheet placed between the source and the GM tube. The measurements are shown in the table. Complete the last column giving your answers to one decimal place.

Absorber (mm)	Number of counts	Counting time (s)	Count-rate in counts per second
1	1601	100	
2	822	100	
3	401	100	
4	399	100	

d The range of beta radiation in aluminium is the minimum thickness needed to stop all the beta particles. Use the data in the table to determine the range of beta radiation in aluminium to the nearest mm. Justify your answer.

Range = ________________ mm

____

____ [3 marks]

2. **a** Complete the table giving the range of alpha, beta and gamma radiation in air.

Radiation	Range
Alpha	
Beta	
Gamma	

[3 marks]

b Suggest which of the radiations produces the most ions per cm of their path through air.

____ [1 mark]

Radioactive decay equations

1. An americium nucleus, $^{241}_{95}Am$, emits an alpha particle and forms a nucleus of neptunium.

Complete the decay equation below by giving the mass number and atomic number of neptunium and also the symbol for an alpha particle.

$^{241}_{95}\text{Americium} = {}^{_}_{_}\text{Neptunium} + $ ________________ [3 marks]

2. The nucleus of aluminium, $^{28}_{13}Al$, decays by emitting a beta particle and forms a nucleus of silicon (Si). Complete the decay equation below by giving the mass number and atomic number of silicon and also the symbol for a beta particle.

$^{28}_{13}Al = {}^{_}_{_}Si + $ ________________ [3 marks]

3. **a** The isotope of magnesium $^{27}_{12}Mg$ is unstable. To become more stable, a neutron in its nucleus changes into a proton and an electron. The daughter nucleus is an isotope of aluminium (Al). Give the name of this type of decay.

________________________________ [1 mark]

b Complete the equation for this decay.

$^{27}_{12}Mg = $ ________________ + ________________ [2 marks]

c The nucleus produced in this decay has an excess of energy. Describe how this nucleus transfers this excess of energy to its surroundings.

____

________________________________ [1 marks]

Half-lives

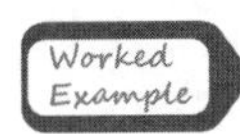

A sample of thorium, $^{234}_{90}Th$, gives a count rate of 360 counts/second. 72 days later the count-rate has fallen to 45 counts/second. Calculate the half-life.

Fraction of thorium remaining after 72 days = $\frac{45}{360} = \frac{1}{8}$

$\frac{1}{8} = \frac{1}{2} \times \frac{1}{2} \times \frac{1}{2}$ so three half-lives have passed.

Therefore, three half-lives = 72 days,

which gives one half-life = 24 days.

Marks gained: [3 marks]

1. Maths Skills Higher Tier only

a The count-rate due to a sample of sodium, $^{24}_{11}Na$, is 1 920 counts/second and 30 hours later the count-rate has fallen to 480 counts/second. Calculate the fraction of the sample of sodium remaining after 30 hours.

Fraction remaining = ______________ [1 mark]

Maths Skills

b Calculate the half-life of the isotope of sodium.

Half-life = ______________ hours [2 marks]

2. Maths Skills Higher Tier only

a The mass of a freshly prepared sample of a radioactive isotope is 1.20 μg. Calculate the mass of the isotope remaining after 48 days given that the half-life is 16 days.

Mass remaining = ______________ μg [3 marks]

Maths Skills Higher Tier only

b The initial activity of the sample is 3 600 Bq. Calculate the activity of the sample after 48 days.

Activity = ______________ Bq [2 marks]

3.

a Iodine $^{131}_{53}I$ decays to form an isotope of xenon $^{131}_{54}Xe$. State what type of decay takes place.

______________ [1 mark]

Maths Skills Higher Tier only

b $^{131}_{53}I$ has a half-life of 8 days. A sample of this isotope is measured as having an activity of 12 kBq. Predict the sample's activity 32 days later. Give your answer in Bq.

Activity = ______________ Bq [3 marks]

4. a The radioactive decay of an unstable nucleus is described as a random event. Describe what is meant by a 'random event'.

______________________________ [1 mark]

Maths Skills

b Tossing a coin is a random event. A student has 10 two pence coins in his hand and throws them up in the air. The coins land on the floor with 7 coins showing heads and 3 showing tails. Calculate the percentage of coins showing heads.

Percentage = ______________% [1 mark]

Maths Skills

c The student puts the 10 coins in a bucket with 190 others and throws all two hundred coins into the air. 96 of the coins land on the floor showing heads and 104 land showing tails. Calculate the percentage of coins showing heads.

Percentage = ______________% [1 mark]

d Suggest what your answers to parts (b) and (c) show about the predicted outcomes in the random event of tossing a coin.

____

____

______________________________ [2 marks]

e In the radioactive decay of an unstable nucleus the number of radioactive atoms in the sample halves as each half-life passes. This pattern is observed even if the mass of the sample is just a few micrograms. Explain why.

______________________________ [1 mark]

5. **Figure 4** is a graph of count-rate versus time for the decay of a radioactive isotope.

Maths Skills

a Draw a line of best fit on the graph. [1 mark]

Maths Skills

b Use the graph to determine two values for the half-life of the radioactive isotope. Clearly show on the graph how you obtained the half-life values. Calculate a mean value for the half-life.

Count-rate (counts/second)

120, 100, 80, 60, 40, 20, 0

0, 50, 100, 150, 200

Time (s)

Figure 4

First half-life value = ______________ s

Second half-life value = ______________ s

Mean half-life = ______________ s [5 marks]

Radioactive contamination

1. **a** Workers at nuclear facilities are at risk from radioactive contamination. Explain what is meant by radioactive contamination.

____

____

______________________________ [1 mark]

b Suggest **three** ways that radioactive atoms could enter a worker's body.

____

____

____

____

______________________________ [3 marks]

c In designated areas at nuclear power stations and nuclear research laboratories, workers are required to wear protective clothing and a dust mask. Give **three** ways the protective clothing and dust mask prevent contamination.

____

____

____

____

______________________________ [3 marks]

2. A radioactive isotope of cobalt, $^{60}_{27}Co$, is used in food irradiation and the irradiation of medical and dental equipment. This isotope of cobalt emits gamma rays. Explain why the wearing of protective clothing would not protect a worker from the gamma rays.

____

______________________________ [1 mark]

Background radiation

1. **Figure 5** shows the main sources of background radiation in the UK. The percentages given are average values. Individual doses of background radiation vary considerably and also depend on location.

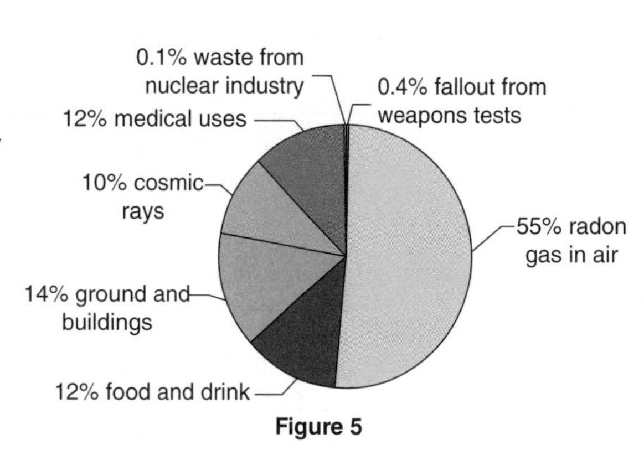

Figure 5

a State the largest natural source of background radiation.

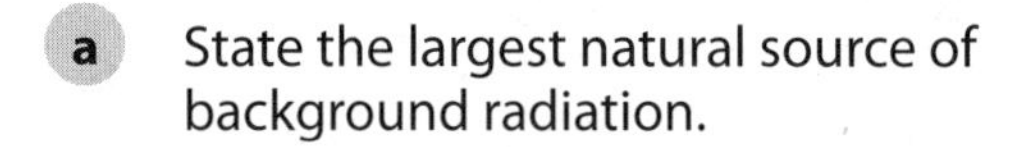

_______________ [1 mark]

b State the largest optional man-made source of background radiation.

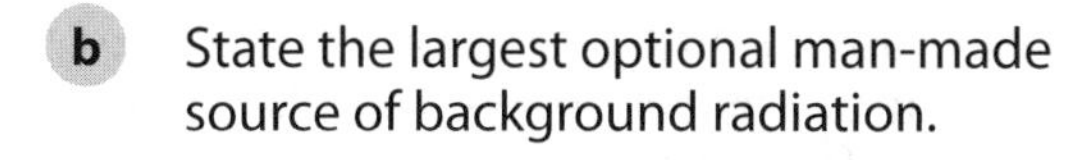

_______________ [1 mark]

c Radon is an alpha emitter and is a gas at room temperature. Explain why radon is a health risk.

____

_______________ [2 marks]

2. Radiation dose is measured in sieverts (Sv). The radiation dose takes account of the energy absorbed and the biological effects of different types of ionising radiation on the human body. Some typical radiation doses are given in the table.

Exposure to background radiation	Radiation dose (mSv)
Average UK annual dose	2.7
Annual dose for residents of Cornwall	7.8
Typical chest X-ray	0.02
Typical flight from the UK to Spain	0.01

a The higher annual exposure to background radiation in Cornwall is due to increased levels of radon, a radioactive gas produced by the decay of radioactive atoms naturally present in certain types of rock. Suggest why the annual radiation dose varies throughout the UK.

____

_______________ [2 marks]

b Suggest why taking a flight increases your radiation dose.

___

___

___ [2 marks]

Uses and hazards of nuclear radiation

1. There are many industrial uses of radioactive isotopes. For example, the location of a leak in an underground oil pipe can be found by mixing a small amount of the radioactive isotope sodium-24 with the oil flowing in the pipe. Sodium-24 emits both beta and gamma radiation. The increased level of radioactivity at the location of the leak is detected using a Geiger–Muller tube.

a Suggest why an alpha emitter would not be suitable for this purpose.

___

___ [1 mark]

b The half-life of sodium-24 is 15 hours. Explain why this value of half-life makes sodium-24 suitable for finding a leak in an underground oil pipe.

___

___

___ [2 marks]

2. Technetium-99m is the most commonly used radioactive isotope used as a medical tracer. Technetium-99m emits only gamma rays and has a half-life of 6 hours. The gamma rays emitted from technetium-99m are lower energy than most gamma ray sources. Explain why technetium-99m is suitable for use as a medical tracer.

___

___

___

___

___

___ [4 marks]

Nuclear fission

1. **a** In a nuclear reactor, fission of some of the uranium-235 atoms in the nuclear fuel is induced by introducing a source of neutrons into the reactor. A chain reaction can be established. Describe what is meant by a chain reaction.

____

____

______________________________ [2 marks]

b The nuclei of the uranium-235 atoms contain a store of energy. The fission reaction releases energy. Describe how the fission products store the energy released by fission.

____

______________________________ [2 marks]

c When a reactor at a nuclear power station is started up, its core temperature rises. Explain how a large number of fission reactions result in an increase in the core's temperature.

____

______________________________ [2 marks]

d In a nuclear reactor the speed of the chain reaction is controlled by raising or lowering control rods. Explain how lowering the control rods affects the speed of the chain reaction.

____

____

______________________________ [2 marks]

2. The fission product isotopes produced by fission of uranium-235 in a nuclear reactor are described as high level waste. The isotopes are radioactive. A sample of high level waste initially contains equal numbers of atoms of the fission product isotopes shown in the table.

Fission product isotope	Radiation emitted	Half-life
Barium-140	beta	13 days
Strontium-90	beta	29 years
Caesium-135	beta	2.3×10^6 years

Describe how the hazard due to the sample varies as time passes.

____

____

____

____

____

____ [4 marks]

Nuclear fusion

1. Inside the Sun, 600 million tonnes of hydrogen are converted into helium every second. This releases a tremendous amount of energy. The temperature at the core of the Sun is 15 million degrees Celsius. At this temperature, the electrons in hydrogen atoms and the protons that form the nuclei of the hydrogen atoms are separated. The electrons and protons move around at very high speeds. Two protons can fuse to form the nucleus of an isotope of hydrogen called a deuteron (**Figure 6**). This is the first stage of the fusion reaction that converts hydrogen to helium inside the Sun.

before after

Figure 6

a For two particles to fuse together they must get very close. Explain why the free protons would have to be moving very fast to get close together.

____

____

____ [2 marks]

b The symbol for a deuteron is $^{2}_{1}H$. Suggest what change has taken place to one of the protons in the fusion reaction shown in **Figure 6** to create a deuteron.

____ [1 mark]

c The second stage of fusion involves a proton undergoing a fusion reaction with a deuteron to form the nucleus of an isotope of helium (**Figure 7**).

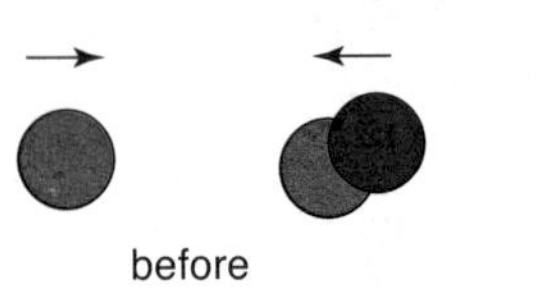

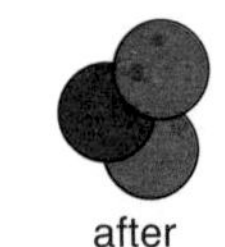

Figure 7

Give the atomic number and the mass number of the nucleus created.

Atomic number = ____________ Mass number = ____________

[2 marks]

d Use your answer to part (c) to help complete the equation for the reaction shown in **Figure 7**.

__hydrogen + __hydrogen = __helium [3 marks]

e The third stage of fusion involves two of the light helium nuclei fusing together to form a heavier nucleus of helium. This reaction releases two free protons (**Figure 8**). Helium is represented by He and hydrogen by H.

before

after

Figure 8

Write the equation for the reaction shown in **Figure 8**.

____

____

____

____

______________________________ [4 marks]

Scalars and vectors

1. In order to go round an island, a boat takes the route from **A** to **B** (**Figure 1**).

a Determine the distance the boat travels from A to B.

N

diagram scale 5 km

A

B

N W E S

Figure 1

Distance = ____________________ km [1 mark]

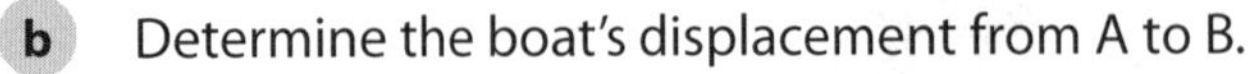

b Determine the boat's displacement from A to B.

Distance = ____________________ km Direction: ____________________ [2 marks]

2. An object's displacement is represented by the the arrow shown in **Figure 2**. Determine the object's displacement.

Maths Skills

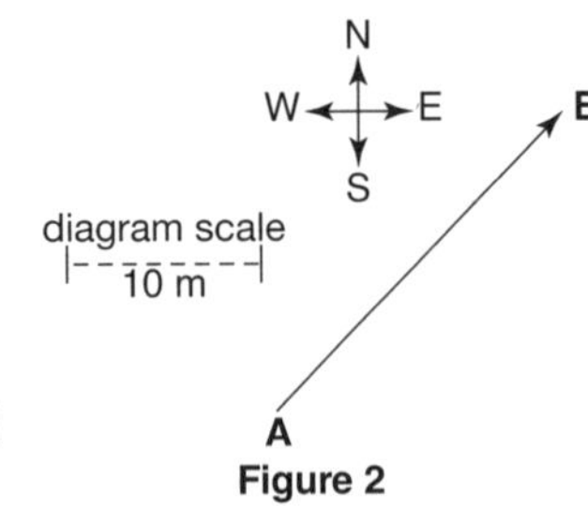

Figure 2

Distance = ____________________ m

Direction: ____________________ [2 marks]

Speed and velocity

1. **a** In an indoor athletics event, the 60 m race is run on a straight track. A world-class sprinter wins the race in 6.4 s. Calculate his average velocity in m/s and in km/h. Give your answers to two significant figures.

Maths Skills

Average velocity = ____________________ m/s

or ____________________ km/h [6 marks]

Maths

You need to learn the equation that links distance travelled to speed and time as you may not be given the equation in the question.

2. Which of the following is a suitable estimate for how long it would take an Olympic cyclist to complete the 4 km pursuit? Tick **one** box.

Maths Skills

☐ 100 s ☐ 300 s ☐ 1000 s ☐ 3000 s

[1 mark]

Remember

It is useful to know these typical speed estimates so you can check that your answers to velocity calculations are realistic: walking 1.5 m/s; jogging 3 m/s; cycling 6 m/s; car 20 m/s; aircraft 200 m/s.

3. The Olympic triathlon involves three consecutive races – a swim, a cycle ride and a run. An athlete's times for the three stages of the race are shown in the table

Maths Skills

a Complete the third column of the table with his average speed values for each stage.

Give each value to two significant figures.

Stage	Time (minutes)	Average speed (m/s)
1500 m swim	25	
40 km cycle	70	
10 km run	42	

[4 marks]

b Calculate his average speed for the **whole** race.
Give your answer to two significant figures.

Average speed = ______________ m/s [3 marks]

c Sketch a distance–time graph for the triathlete on the axes in **Figure 3**. [3 marks]

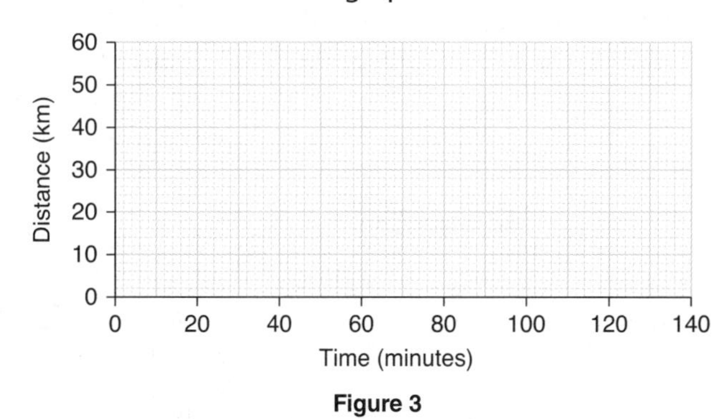

Figure 3

4. The displacement–time graph for a car on a straight test track is shown in **Figure 4**.

Maths Skills

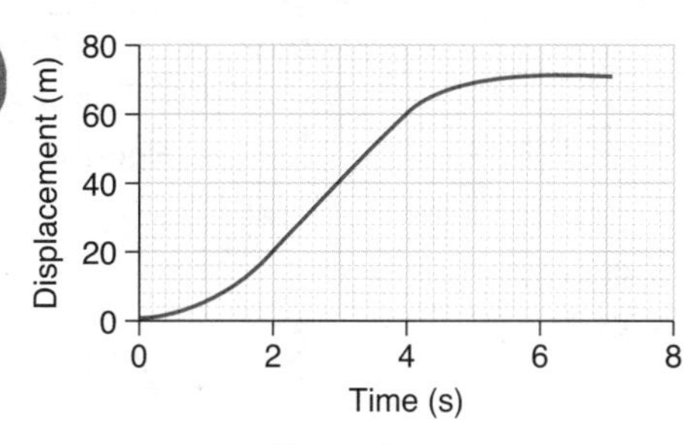

Figure 4

Maths

It is important to know that the gradient of an object's displacement–time graph is equal to its velocity.

Higher Tier only

Determine the car's velocity at 3 s. Show clearly how you work out your answer.

Velocity = ______________ m/s [3 marks]

Acceleration

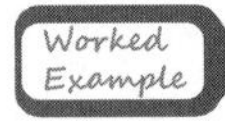

A car is travelling at a steady speed of 36 km/h. The car then accelerates for 5.0 s until it reaches a speed of 54 km/h. Calculate the car's acceleration in m/s^2.

Converting speeds in km/h to m/s: $36 \text{ km/h} = 36 \times \frac{1000 \text{ m}}{3600 \text{ s}}$

$= 10 \text{ m/s}$

$54 \text{ km/h} = 54 \times \frac{1000 \text{ m}}{3600 \text{ s}} = 15 \text{ m/s}$

acceleration $a = \frac{\Delta v}{t}$

$= \frac{15 - 10}{5.0} = 1.0 \text{ m/s}^2$

Marks gained: [3 marks]

Maths

You need to know the equation that relates acceleration to velocity change and time taken as you may not be given the equation in the question.

1. Maths Skills

A cyclist is travelling along a straight road at a steady speed of 22 km/h. He accelerates for 8.0 s and reaches a speed of 28 km/h. Calculate his acceleration in m/s^2. Give your answer to two significant figures.

Maths

You need to know that the gradient of a velocity–time graph gives the acceleration.

Acceleration = ______________ m/s^2 [5 marks]

2. Maths Skills

Figure 5 shows the velocity–time graph for a car for its first 4 s of motion.

Calculate the car's acceleration. Show clearly how you work out your answer.

Velocity (m/s): 0, 5, 10, 15, 20

Time (s): 0, 1, 2, 3, 4

Figure 5

Acceleration = ______________ m/s^2 [2 marks]

3. Maths Skills

Higher Tier only

The graph in **Figure 6** shows how the velocity of a sky diver changes from the moment she jumps out of the aircraft until her parachute opens.

Use the graph to estimate the sky diver's acceleration 10 s into the dive. Show clearly how you work out your answer.

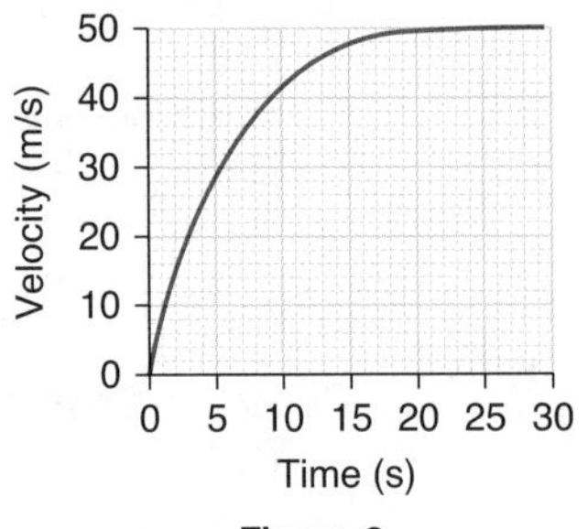

Figure 6

Acceleration = ______________ m/s^2 [3 marks]

4. Higher Tier only

Figure 7 is a velocity–time graph for a car moving along a straight test track.

Calculate the car's displacement during the 20 s of motion. Show clearly how you work out your answer.

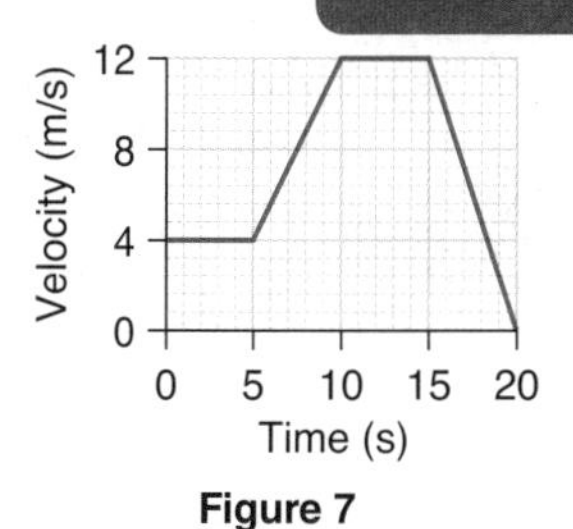

Figure 7

Displacement = ______________ m [3 marks]

Maths
You need to know that the area between the line of a velocity–time graph and the time axis gives the displacement.

Equation for uniform acceleration

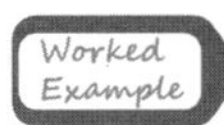

A cyclist accelerates from a stationary position along a straight road with an acceleration of 0.40 m/s^2. Calculate his velocity after he has travelled a distance of 45 m.

First match the data to the equation symbols: $u = 0$, $a = 0.40$, $s = 45$.

Now substitute the data into the equation:

$$v^2 - u^2 = 2as$$

$$v^2 - 0 = 2 \times 0.40 \times 45$$

which gives $v^2 = 36$ so his velocity is 6.0 m/s.

Marks gained: [2 marks]

Maths
The equation $v^2 - u^2 = 2as$ is on the Physics Equations Sheet. You are expected to be able to select and apply this equation.

1. Maths Skills

A coin is dropped from Blackpool Tower and falls to the ground, a distance of 150 m. Assuming that air resistance has no effect on the coin, calculate the speed at which it would hit the ground. Give your answer to two significant figures.

Acceleration due to gravity = 9.8 m/s^2.

Velocity = ______________ m/s [4 marks]

2. Maths Skills

An electric tram moving with a velocity of 4.0 m/s comes to a 25 m long straight section of track and accelerates with an acceleration of 0.40 m/s². Calculate the tram's velocity when it has travelled the 25 m along this straight section of track. Use the correct equation from the Physics Equations Sheet.

Tram's velocity = ____________ m/s [3 marks]

3. Maths Skills

An astronaut on the Moon releases an object from a height of 2.0 m. Its velocity on hitting the ground is measured as 2.5 m/s. Calculate the acceleration due to gravity on the Moon. Give your answer to two significant figures.

Acceleration = ____________ m/s^2 [4 marks]

Forces

Worked Example

Figure 8 shows a football stationary on the ground. Add labelled arrows to the diagram to show the forces acting on the football.

Remember

'Normal' means at right angles.

Answer

normal contact force

force of gravity

Figure 8

Marks gained: [2 marks]

1. **a** **Figure 9** shows a falling leaf.

Add labelled arrows to the diagram to show the forces acting on the leaf. [2 marks]

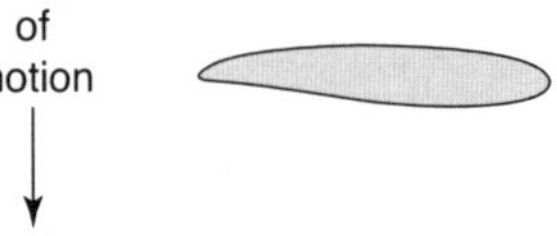

direction of motion

Figure 9

b State whether the forces you have added to **Figure 9** are contact or non-contact forces.

____

______________________________ [2 marks]

2. A student applies a small force to the newton meter attached to a block of wood positioned on a table (**Figure 10**). However, the block does not move.

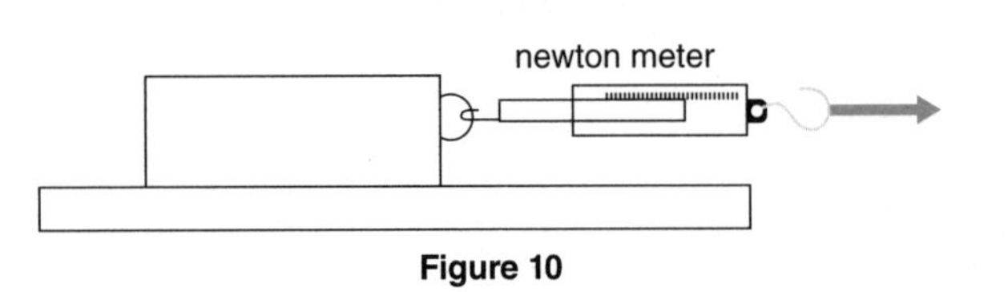

Figure 10

On the diagram add labelled arrows to show the forces acting on the block in addition to the force exerted by the newton meter. [3 marks]

Moment of a force

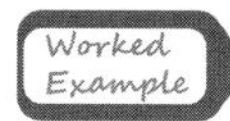

A person closing a gate exerts a force of 5 N (**Figure 11**). The line of action of the force is shown by the red line. Calculate the moment exerted by the person closing the gate. Give the answer in N m.

The perpendicular distance from the line of action of the force to the hinge is 60 cm.

Moment = Fd

= 5 N × 0.6 m

= 3 N m

Maths

You need to learn the equation relating the moment of a force to the size of the force and the perpendicular distance from the line of action of the force to the pivot.

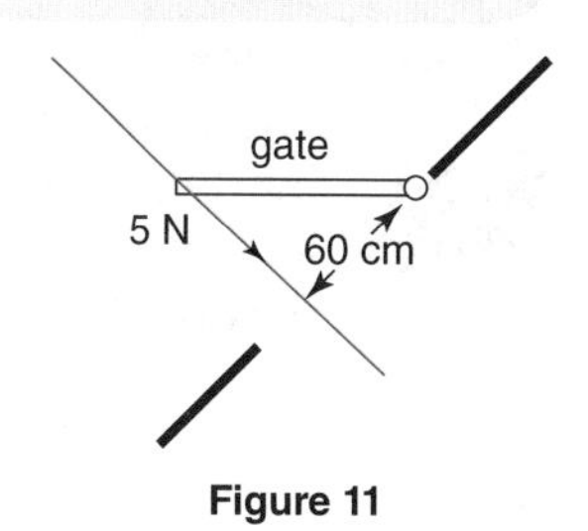

Figure 11

Marks gained: [2 marks]

1. Maths Skills

A person exerts a force on a door handle (**Figure 12**). The door has dimensions of 76 cm by 200 cm. The force exerted on the door handle is 10 N.

Calculate the moment of this force about the hinge. Give your answer in N m.

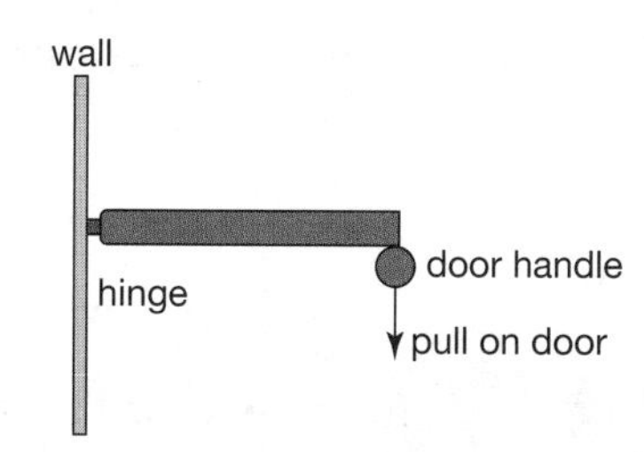

Figure 12

Moment = ________________ N m [2 marks]

2. Maths Skills

A metre rule is pivoted on a nail through a hole at its centre. The rule supports two weights and is balanced horizontally as shown in **Figure 13**.

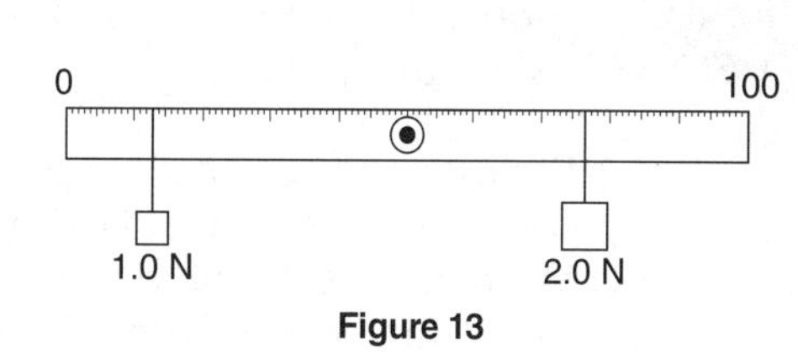

Figure 13

a The 1.0 N weight is positioned at the 12 cm mark of the metre rule. Calculate the moment exerted by this weight about the pivot. Give your answer in N cm.

Moment = ________________ N cm [2 marks]

b Calculate the distance from the pivot to the line of action of the force exerted by the 2.0 N weight on the metre rule. Give your answer in cm.

Distance = ________________ cm [3 marks]

c At what position on the metre rule is the 2.0 N weight supported? Give your answer in cm from the 0 cm mark on the rule.

cm mark on rule: ________________ [1 mark]

Levers and gears

1. Maths Skills

A gardener wants to move a boulder. The weight of the boulder is 1000 N. He decides to use a plank of wood as a lever to make lifting the boulder easier (**Figure 14**).

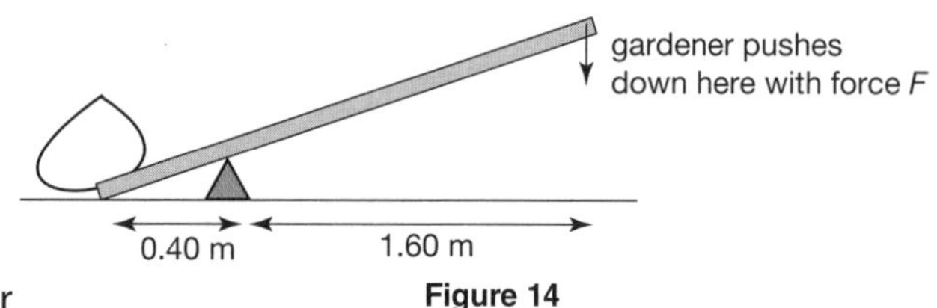

Figure 14

a Calculate the force that the gardener would have to exert in order to just lift the boulder.

Force = ________________ N [3 marks]

b The biggest force that the gardener can exert is 360 N. Calculate the maximum weight the gardener can lift using the lever set up as shown in **Figure 14**. Give your answer in kN to two significant figures.

Maximum weight = ________________ kN [4 marks]

2. Maths Skills

To move some soil, a gardener has to lift the handle of the wheelbarrow (**Figure 15**) so that the back legs are just off the ground.

The biggest force that the gardener can exert is 360 N. Calculate the maximum combined weight of the soil and wheelbarrow that the gardener can lift. Give your answer in kN to two significant figures.

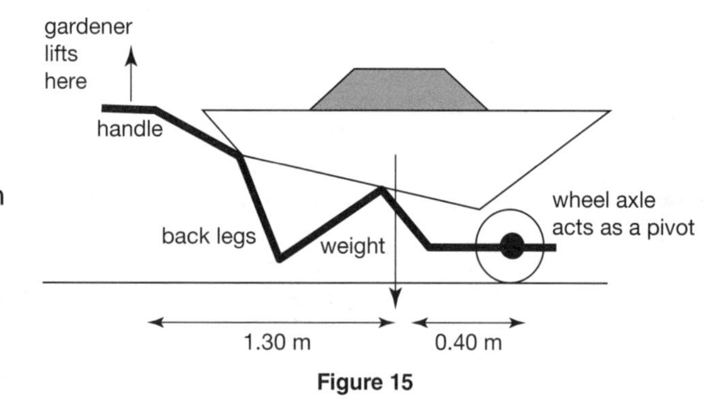

Figure 15

Maximum weight = ________________ kN [4 marks]

3. Levers and gears are used to transmit rotational forces.

In **Figure 16**, the larger gear is the driver gear and the smaller gear is the driven gear. State which gear completes more revolutions per second.

______________________ [1 mark]

Figure 16

Pressure in a fluid

1. Pressure is usually measured in pascals (Pa). Use the equation that relates pressure to the normal force and area to obtain alternative units that are equivalent to Pa.

____ [2 marks]

> **Maths**
> You need to know the equation that relates pressure to normal force and area as you may not be given this equation in the exam.

2. Maths Skills

a Describe the direction of the force exerted by a liquid or gas on an surface with which it is in contact.

____ [1 mark]

b The surface area of an average 10-year-old child is 1.14 m^2. At sea level the pressure due to the atmosphere is 1.01×10^5 Pa. Calculate the total force exerted by the atmosphere on the child's body at sea level. Give your answer in standard form to three significant figures.

Force = ____________________ N [4 marks]

> **Maths**
> The equation for the pressure due to a column of liquid is on the Physics Equations Sheet. You need to be able to select and apply this equation.

3. Maths Skills — Higher Tier only

A scuba diver is 50 m below the surface of the ocean. Calculate the pressure on the diver due to the sea water. Select the correct equation from the Physics Equations Sheet.

Density of seawater = 1.0×10^3 kg/m^3.

Gravitational field strength = 9.8 N/kg.

Pressure = ____________________ Pa [3 marks]

Atmospheric pressure

1. Maths Skills

Atmospheric pressure varies with height above sea level (**Figure 17**). Annapurna, a mountain in the Himalayas, is about 8000 m above sea level.

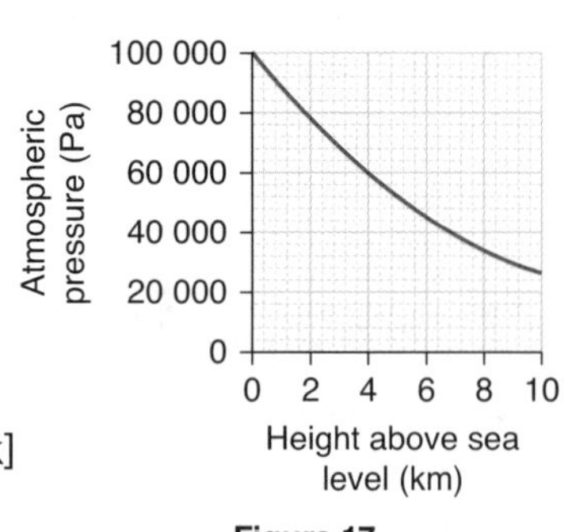

Figure 17

a Use **Figure 17** to estimate the value of atmospheric pressure at the summit. Show on the graph how you obtained the answer.

Pressure = ____________________ Pa [1 mark]

b Some aircraft use a pressure altimeter to indicate their height above sea level. Use **Figure 17** to predict the altitude of an aircraft if the atmospheric pressure was measured at 60 kPa. Show on the graph how you obtained the answer.

Altitude = ____________________ m [1 mark]

Literacy

A 2-mark written question expects you to make **two** points. Similarly, 3 marks means **three** points are required.

2. **a** Explain, with reference to molecules in the air, how the atmosphere exerts pressure on a surface.

____

____

____ [2 marks]

b **Figure 18** shows a metal can before and after being evacuated by a vacuum pump.

Explain, with reference to gas molecules, the change in shape of the can.

to vacuum pump

metal can

before

after

Figure 18

____

____

____

____

____ [3 marks]

Gravity and weight

1. The gravitational field strength at the Earth's surface is 9.8 N/kg and on the surface of the Moon it is 1.6 N/kg. Describe how the mass and weight of a particular astronaut on the Earth would compare with his mass and weight on the Moon.

Maths
You need to know the equation that relates the weight of an object to the object's mass and the gravitational field strength in the location of the object.

____

____

____

______________________________ [2 marks]

2. Synoptic Maths Skills

The fuel tank of a long distance lorry has a volume of 0.5 m^3. At the start of a journey the tank is full. If all the fuel is used on a journey, calculate the change in the lorry's weight. Write down the equations that you use.

Density of fuel = 800 kg/m^3. Gravitational field strength = 9.8 N/kg.

Change in weight = ______________ N [5 marks]

Resultant forces and Newton's first law

1. Maths Skills

Figure 19 shows a child's trolley and the forces acting on it.

child's trolley

Figure 19

One child exerts a force of 5.1 N in one direction and a second child exerts a force of 11.8 N in the opposite direction. Calculate the resultant force. State its direction.

Resultant force = ______________ N

Direction ______________ [2 marks]

2. **a** State the name that Newton gave to the tendency of an object to resist a change in its motion.

_______________ [1 mark]

b A raindrop falls at steady speed. Add labelled arrows to the raindrop in **Figure 20** to show the relative size and direction of the forces acting on the raindrop. [3 marks]

Figure 20

3. Maths Skills — Higher Tier only

An object is acted on by two forces each of 5.0 N. One force acts downwards and the other force acts to the right. In the box (**Figure 21)** draw a vector diagram to determine the magnitude and direction of the resultant force. [4 marks]

Figure 21

4. Maths Skills — Higher Tier only

In **Figure 22** the arrow represents a 10.0 N force acting on an object. The diagram is drawn to scale.

a Add arrows to the diagram to show the horizontal and vertical components of the force. [1 mark]

b Determine the magnitude of the horizontal and vertical components of the force.

Horizontal component = _______________ N;

vertical component = _______________ N [2 marks]

Figure 22

Forces and acceleration

1. Required practical

A student uses an air track with glider and light gates (**Figure 23**) to investigate the relation between acceleration and the force that causes the acceleration.

When the glider is released at the left end of the air track, it passes through both light gates.

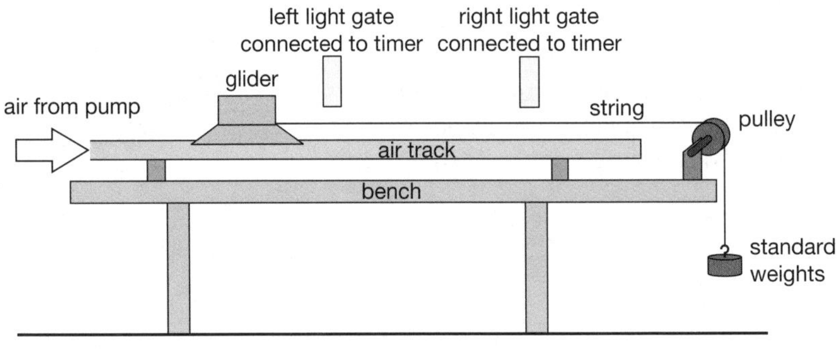

Figure 23

a Describe how the effects of friction and air resistance on the glider are minimised in this experiment.

____

____ [2 marks]

b Describe how the student could check if the glider travels at a steady speed along the air track when there is no accelerating force present.

____

____

____ [2 marks]

c If the friction between the glider and the air track is not zero, the glider will slow down as it travels having been given an initial push. Describe how the apparatus can be adjusted to ensure that the glider travels along the air track at a steady speed when there is no accelerating force present.

____

____ [1 mark]

d The weights attached by string to the glider provide the force that accelerates the glider. The table contains one data set from the experiment.

Force acting on glider (N)	Card length (m)	Distance between gates (m)	Time to pass through first gate (s)	Time to pass through second gate (s)
2.0×10^{-2}	0.100	0.500	1.78	0.52

Use the data to calculate the glider's acceleration. Write down the equations that you use.

Acceleration = ______________________ m/s^2 [6 marks]

Problem solving

In a long calculation question there is usually more than one problem to solve to get the information needed to calculate the final answer. Always show your working so that you get some marks even if your final answer is wrong.

e The student plots a graph of acceleration versus force acting on the glider. Describe the graph that would be obtained if the data showed that acceleration was directly proportional to the force acting on the glider.

Maths

You need to know the equation that links resultant force to mass and acceleration.

____

____ [2 marks]

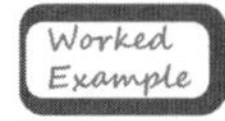

The engine driving force on a car is 2200 N. The resistive forces (friction and air resistance) add up to 500 N. The mass of the car is 850 kg. Calculate the car's acceleration.

The resultant force on the car = 2200 − 500 = 1700 N

$F = ma$ so $a = F \div m$

$= 1700 \div 850 = 2.0\ m/s^2$

Marks gained: [3 marks]

2. Maths Skills

The engine driving force on a truck is 4500 N. The resistive forces (friction and air resistance) add up to 1800 N. The mass of the truck is 2 400 kg. Calculate the truck's acceleration. Give your answer to two significant figures.

Acceleration = ______________ m/s² [4 marks]

3. Maths Skills

a A cyclist is travelling along a straight section of road at a steady velocity of 4.0 m/s. The cyclist and cycle have a combined mass of 63 kg. He then accelerates over a distance of 40 m reaching a speed of 8.2 m/s. Calculate the cyclist's acceleration. Give your answer to two significant figures.

Acceleration = ______________ m/s² [4 marks]

b The resistive forces on the cyclist, air resistance and rolling resistance, add up to 20 N. Calculate the driving force generated by the cyclist.

Driving force = ______________ N [3 marks]

4. The rolling resistance between a bicycle's tyres and the road is relatively constant. However, air resistance increases with speed (**Figure 24**).

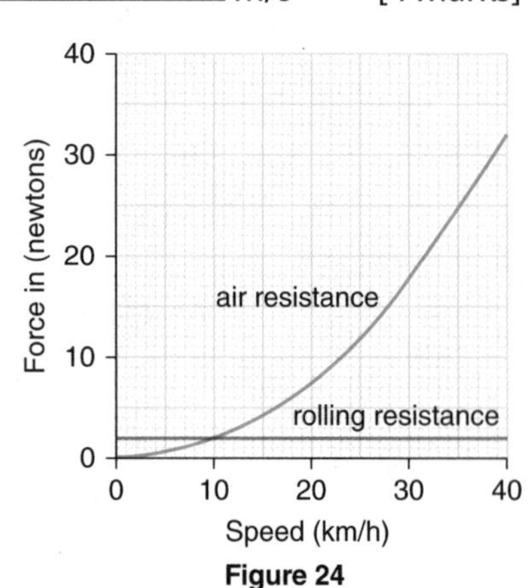

Figure 24

a Determine the speed at which the rolling resistance and the air resistance are the same size.

Speed = ______________ km/h [1 mark]

b A cyclist is cycling along a flat road at a constant speed of 25 km/h. Determine the driving force that the cyclist has to generate to maintain this speed.

Driving force = ______________ N [1 mark]

Terminal velocity

1. A steel ball is released at the surface of the oil in a glass tube (**Figure 25**). The graph shows how the velocity of the ball changes with time.

Maths

The gradient of a velocity–time graph gives the acceleration.

a Describe how the **acceleration** of the ball changes with time.

______________ [1 mark]

b Give the value of the ball's acceleration at position C.

Acceleration = ______________ [1 mark]

c On **Figure 26**, draw free body diagrams showing the weight (W) and the drag (D), acting on the ball corresponding to positions **A**, **B** and **C**. Label the arrows with W for weight and D for drag. [4 marks]

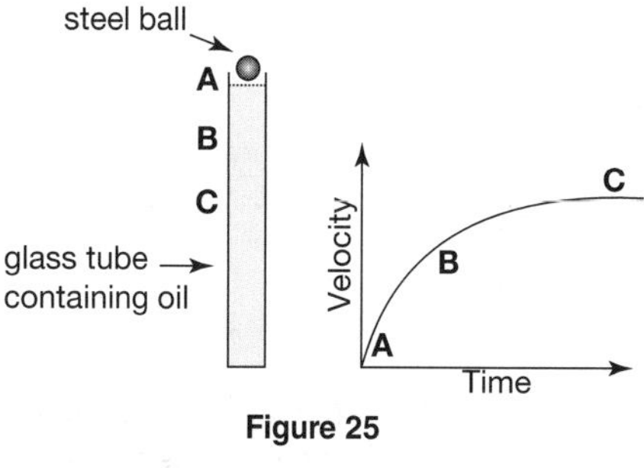

Figure 25

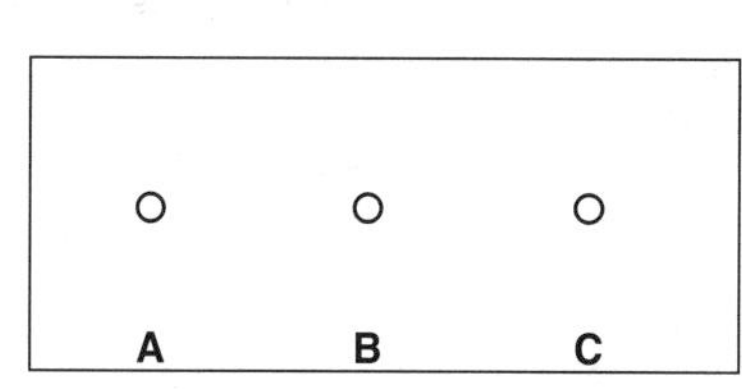

Figure 26

2. **Figure 27** shows the change in the speed of a freefall parachutist before her parachute is opened.

Maths Skills

Higher Tier only

a Use the graph to estimate the parachutist's acceleration at 10 s. Show clearly how you work out your answer.

Acceleration = ______________ m/s^2. [2 marks]

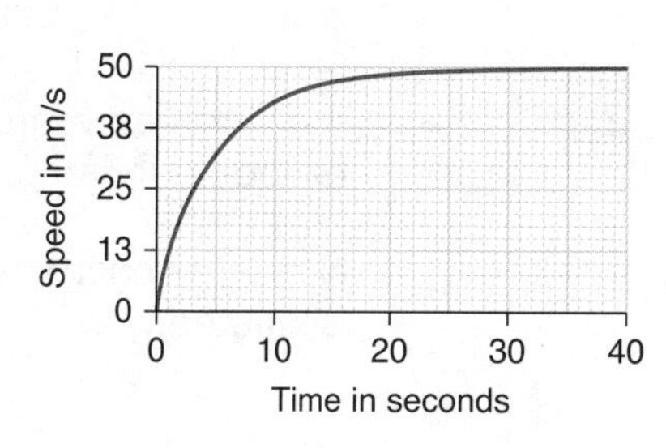

Figure 27

b The mass of the parachutist is 70 kg. Calculate the air resistance acting on her at 10 s. Write down the equations that you use. Gravitational field strength = 9.8 N/kg.

Air resistance = ______________ N [6 marks]

Newton's third law

1. **Figure 28** shows the forces acting on a stationary box on the floor.

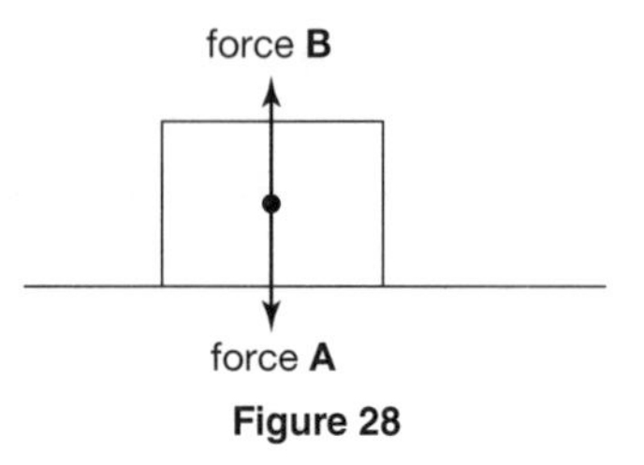

Figure 28

a Describe the force that forms a Newton's third law pair with force **A**. Include the direction, the type of force and the object that the force is exerted on.

____

____

______________ [3 marks]

b Describe the force that forms a Newton's third law pair with force **B**. Include the direction, the type of force and the object that the force is exerted on.

____

____

______________ [3 marks]

2. **a** A balloon powered toy cart is shown in **Figure 29**. When the balloon is inflated, the rubber is stretched.

Name the energy store that is increased when the balloon is inflated.

______________ [1 mark]

b The inflated balloon is released and the toy cart accelerates. Explain why the toy cart accelerates.

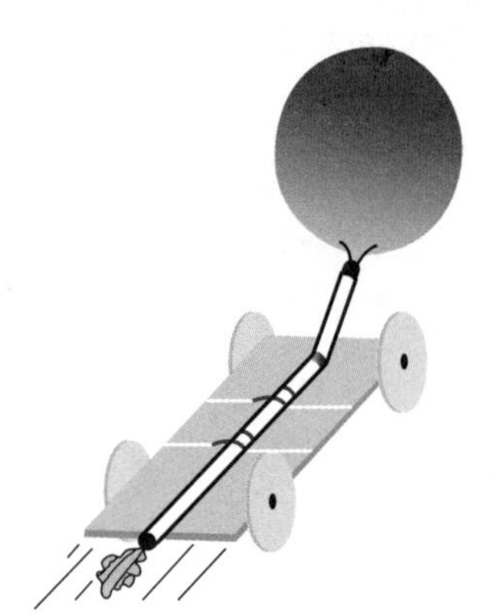
Figure 29

[2 marks]

Work done and energy transfer

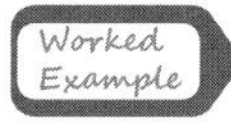

The winch in **Figure 30** drags the 200 kg load up the length of the ramp at a steady speed in 30 s.

Calculate the useful output power of the winch.

Work done = Fs

= 1000 N × 15 m

= 15 000 J

Useful output power = $\frac{W}{t}$

= $\frac{15\,000\text{ J}}{30\text{ s}}$ = 500 W

Maths
You need to know the equation that relates work done to force and distance moved in the direction of the force.

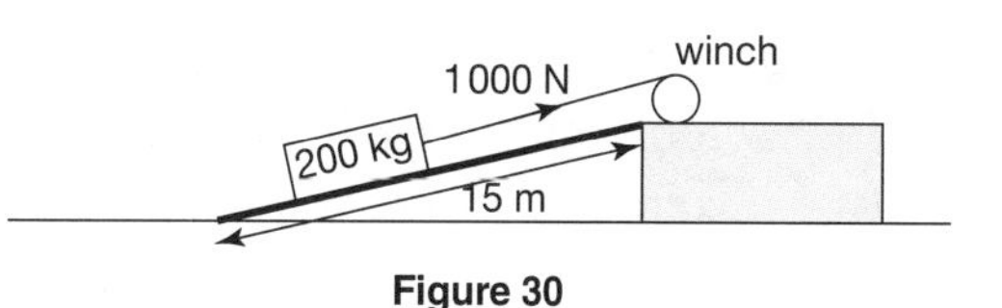

Figure 30

Marks gained: [3 marks]

1. The battery driven electric motor (**Figure 31**) raises a 500 g mass a distance of 100 cm in 2.1 s at a steady speed.

Remember
To lift a mass at a steady speed (no acceleration) requires a force equal to the weight of the mass.

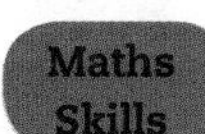

a Calculate the work done by the motor.

Gravitational field strength = 9.8 N/kg.

Work = ____________ J [2 marks]

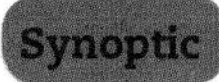

b Calculate the useful output power of the motor. Give your answer to two significant figures.

Power = ____________ W [3 marks]

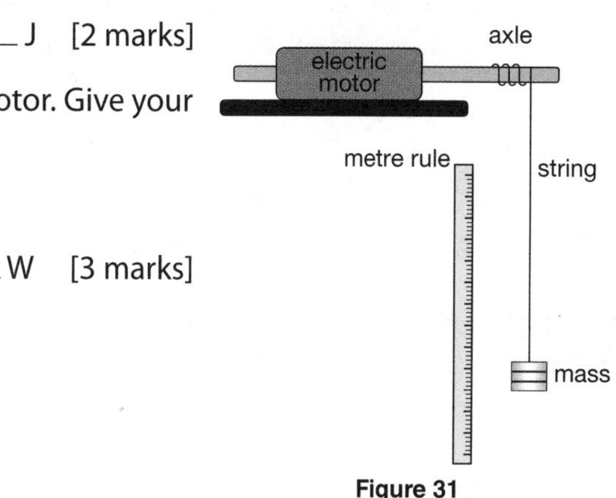

Figure 31

2. A commercial jet aircraft flies at a constant velocity of 800 km/h. The thrust generated by the jet engines is 50 kN.

Maths Skills

a Calculate the work done by the engine when the aircraft flies a distance of 800 km. Give your answer in GJ.

Work = ____________________ GJ [3 marks]

Synoptic

b Calculate the useful output power of the engines. Give your answer in MW to two significant figures.

Power = ____________________ MW [4 marks]

Stopping distance

A car of mass 800 kg is being driven at a steady speed of 10 m/s. The driver sees a hazard ahead and applies the brakes. His reaction time is 0.50 s. The brakes generate a braking force of 5000 N. Calculate the car's stopping distance.

Thinking distance = speed × time = 10 × 0.5 = 5.0 m

Work done = kinetic energy transferred = $\frac{1}{2}mv^2$

$= 0.5 \times 800 \times 10^2$

$= 4.0 \times 10^4$ J

Braking distance = $\frac{\text{work done}}{\text{braking force}}$

$= \frac{4.0 \times 10^4}{5000} = 8.0$ m

Stopping distance = thinking distance + braking distance

$= 5.0 + 8.0 = 13.0$ m

Marks gained: [6 marks]

1. A car is being driven at 72 km/h. The mass of the car including the driver is 800 kg.

Maths Skills

a Calculate the kinetic energy of the car.

Kinetic energy = ____________________ J [3 marks]

b The driver sees a hazard in the road ahead and realises that he has to stop the car. Calculate the thinking distance if the driver's reaction time is 0.60 s.

Thinking distance = ____________________ m [2 marks]

c The car's braking system creates a braking force of 5 000 N. Calculate the distance over which the braking force must act to stop the car. Write down the equations that you use.

Braking distance = ____________________ m [3 marks]

d Calculate the car's total stopping distance.

Stopping distance = ____________________ m [1 mark]

2. The Highway Code for road users in the UK gives a driver's thinking distance while driving a car travelling at 32 km/h as 6.0 m.

Maths Skills

a Calculate the value of the reaction time that has been used to produce this data.

Reaction time = ____________________ s [4 marks]

b Give **two** factors that could affect a driver's thinking time.

1 ____________________ 2 ____________________ [2 marks]

3. Maths Skills

a The graph in **Figure 32** shows the thinking and braking distance for a typical car at various speeds.

Use **Figure 32** to determine the stopping distance for a car travelling at 30 m/s.

Show on the graph how you obtained the answer.

Stopping distance ____________________ m

[2 marks]

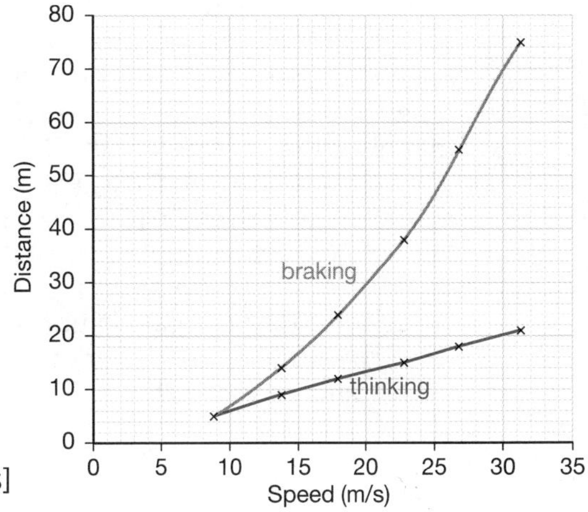

b The car has a mass of 810 kg. Calculate the braking force. Write down the equations that you use.

Braking force = ____________ N [4 marks]

4. Maths Skills

There can be up to about 40% less friction between car tyres on a wet road compared with a dry road. The tread on a tyre is designed to channel water away from the tyre's surface when driving on a wet road. **Figure 33** shows how stopping distances on wet roads are affected as tyres become worn and the depth of the tread decreases.

Determine the percentage increase in stopping distance for a car that has tyres with a 1.6 mm tread (the legal minimum) and the same car that has tyres with a 6.7 mm thread on an asphalt road. Show on the graph how you obtained the answer.

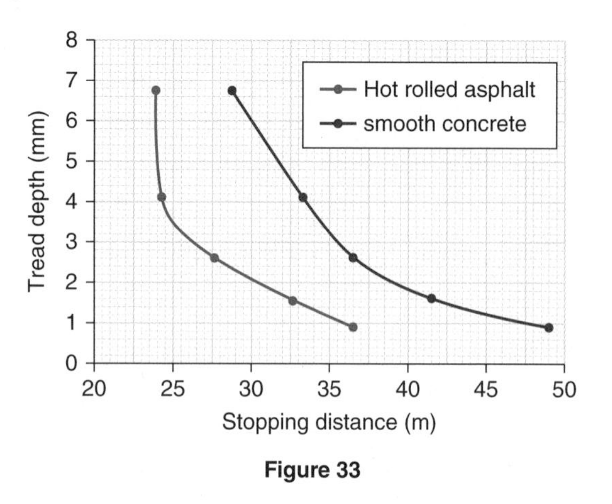

Figure 33

Percentage increase = ____________ % [3 marks]

Force and extension

1. Required practical

A student is given the apparatus shown in **Figure 34** to investigate the relationship between extension and stretching force for a spring. The metre rule, which has been clamped vertically, is positioned so that a pointer taped to the spring points to a value on the metre rule that gives the spring's length. The apparatus includes a set of weights from 1.0 to 8.0 N. The student is instructed not to attach more than 8.0 N to the spring to ensure that it is not elastically deformed.

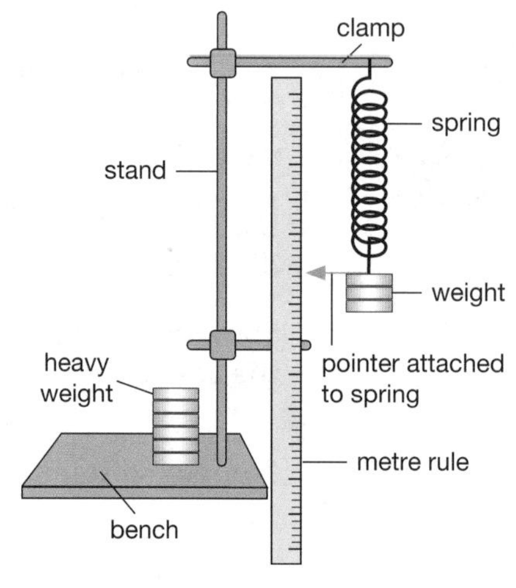

Figure 34

a Add to the diagram in figure 34 to show how the student could check that the rule was clamped vertically.

[2 marks]

b Design a method that the student could follow to investigate the relationship between extension and stretching force. Include a procedure to minimise errors due to random fluctuations and describe how the data produced should be analysed.

Literacy

Check that what you have written makes sense. You cannot get full marks if you contradict your own statements.

[6 marks]

2. A graph of extension versus stretching force for a spring is shown in **Figure 35**.

Maths Skills

a State the range of stretching force values for which the relationship between extension and stretching force is linear.

Range = ______ N [1 mark]

b Determine the spring constant. Show on the graph how you obtained the answer. Give a suitable unit with your answer.

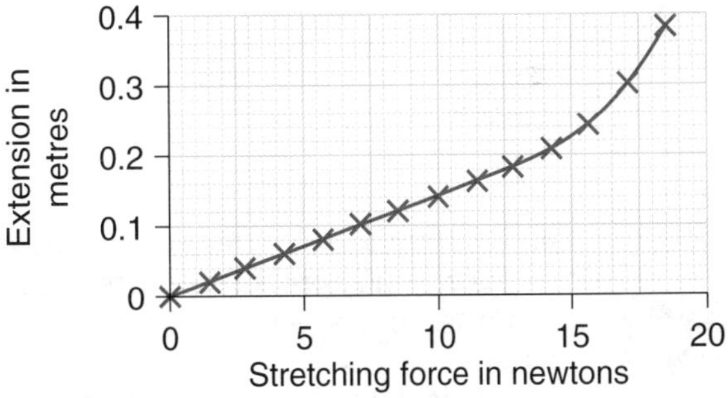

Figure 35

Spring constant = ______ Unit: ______ [5 marks]

c Calculate the elastic potential energy stored in the spring when the stretching force is 10 N. Select the required equation from the Physics Equations Sheet.

Elastic potential energy = ______________ J [2 marks]

3. Maths Skills

Figure 36 shows a simplified diagram of a toy that fires a ping pong ball vertically upwards. The mechanism contains a spring with a spring constant of 20 N/m. When the ball is loaded into the mechanism, the spring is compressed by 10 cm.

Calculate the maximum height reached by the ball above the toy stating any assumption you make. The mass of the ping pong ball is 3.0 g. Give your answer to two significant figures.

Gravitational field strength = 9.8 N/kg.

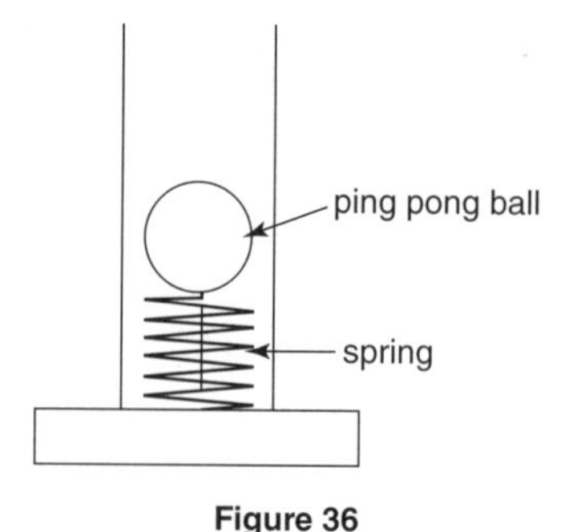

Figure 36

Height = ______________ m [7 marks]

Momentum

1. The velocity–time graph for a 3000 kg lorry on a straight test track is shown in **Figure 37**. Calculate the lorry's change in momentum between $t = 5$ s and $t = 10$ s. Give your answer in standard form to two significant figures.

Maths

You need to learn the equation for momentum as it may not be given to you in the exam.

Momentum change = ______________ kg m/s [5 marks]

2. Maths Skills

a A tennis ball of mass 57.0 g moves horizontally towards a wall (**Figure 38**) and bounces back horizontally at the same speed.

Calculate the ball's momentum as it hits the wall.

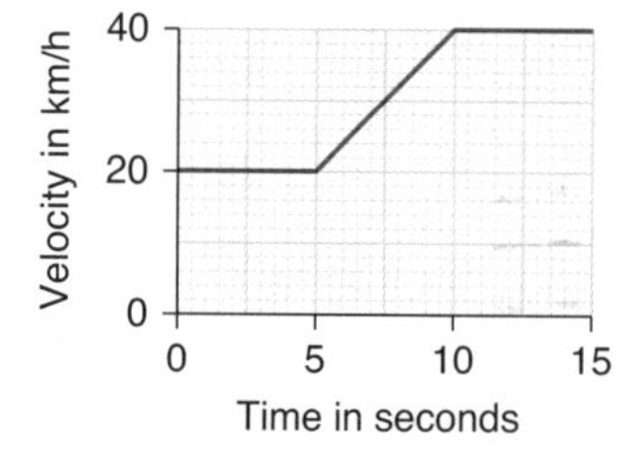

Figure 37

Momentum = ______________ kg m/s [2 marks]

b Calculate the ball's momentum as it bounces back off the wall.

Momentum = ______________ kg m/s [2 marks]

15.0 m/s

Figure 38

c Calculate the magnitude of the ball's momentum change.

Momentum change = ______________________ kg m/s [1 mark]

Conservation of momentum

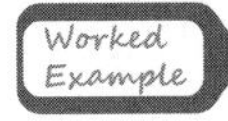

Figure 39 shows two fairground dodgem cars about to collide. After the collision, the blue car has a velocity of 2.0 m/s. Calculate the velocity of the red car after the collision.

100 kg 2.0 m/s → 80 kg 1.0 m/s →

before collision

Figure 39

Momentum before collision = (100 kg × 2 m/s) + (80 kg × 1 m/s) = 280 kg m/s

Momentum after the collision = momentum before collision

$100 \text{ kg} \times v_{red} + (80 \text{ kg} \times 2 \text{ m/s}) = 280 \text{ kg m/s}$

Velocity of the red car: $v_{red} = \frac{280 - 160}{100}$

$= 1.2 \text{ m/s}$

Marks gained: [5 marks]

1. **Maths Skills** A car of mass 800 kg, moving at a speed of 10 m/s collides into the rear of a 1000 kg van moving at 6.0 m/s. After the collision, the vehicles separate and the van has a velocity of 8.0 m/s. Calculate the velocity of the car after the collision. Write down the equations that you use.

Velocity = ______________________ m/s [5 marks]

2. **Maths Skills** **a** A subatomic collision occurs between a neutron and a stationary boron nucleus. The neutron is moving at a velocity of 2.70×10^{7} m/s. The mass of the neutron is 1.70×10^{-27} kg and the mass of the boron nucleus is 1.66×10^{-26} kg. Calculate the total momentum before the collision.

Momentum = ______________________ kg m/s [2 marks]

b In the collision, the neutron is absorbed by the boron nucleus. Their combined mass is 1.83×10^{-26} kg. Calculate the velocity of this combined nucleus. Give your answer to three significant figures.

Velocity = ______________ m/s [5 marks]

Rate of change of momentum

A high jumper of mass 60 kg clears the high jump bar and lands on the crash mat with a speed of 2.8 m/s. The mat brings her to a stop in 0.40 s. Calculate the impact force on the high jumper's body.

$$\text{Impact force} = \frac{m\Delta v}{\Delta t} = \frac{60 \times 2.8}{0.40} = 420\,\text{N}$$

Marks gained: [2 marks]

Maths

The equation $F = \frac{m\Delta v}{\Delta t}$ is on the Physics Equations Sheet. You are expected to be able to select and apply this equation.

1. Car seat belts protect people in two ways during a crash. Firstly they prevent the occupants of the car being thrown about and possibly hitting the windscreen or other sections of the car's interior. Seat belts are also designed to stretch a bit during the impact so that the occupants of the car come to a stop in a longer time than the car does.

a A car being driven at a speed of 15 m/s is involved in a collision. The driver has a mass of 60 kg. He is **not** wearing a seat belt and is stopped by the windscreen in 0.05 s. Calculate the impact force on the driver. Use the correct equation from the Physics Equations Sheet.

Force = ______________ N [2 marks]

b Had the driver been wearing a seat belt he would have been brought to a stop in 0.6 s. Calculate the impact force on the driver.

Force = ______________ N [2 marks]

2. Acceleration sensors in a car trigger an air bag to inflate during a collision. As the driver's head hits the air bag, the bag is squeezed and the nitrogen gas inside escapes slowly through a hole in the bag. Explain, with reference to momentum change, how the air bag protects the driver from injury.

____

____

____

____

____

____ [4 mark]

Transverse and longitudinal waves

1. Waves on a water surface can be produced by a beam vibrating up and down on the water surface. **Figure 1** represents a snapshot of a surface wave travelling in a tank of water.

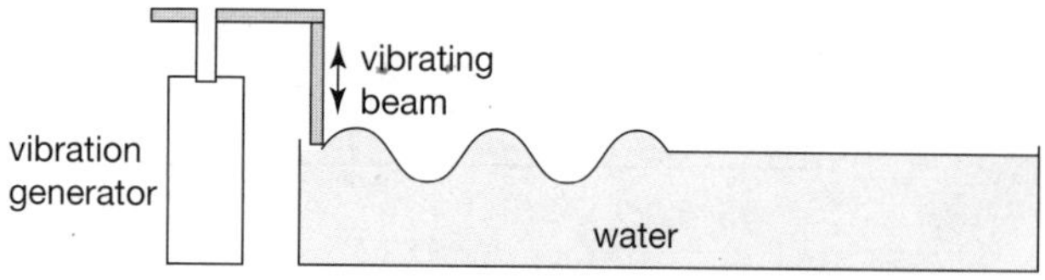

vibrating beam
vibration generator
water

Figure 1

a Describe how the apparatus in **Figure 1** could be used to demonstrate that the water surface only moves up and down even though the wave is travelling to the right.

____

____

____ [2 marks]

Maths Skills **b** Give the angle between the direction of vibration of the water surface and the direction of energy transfer.

____ [1 mark]

2. **Figure 2** represents a snapshot of a longitudinal wave travelling on a slinky spring. Each short vertical line represents a turn on the slinky.

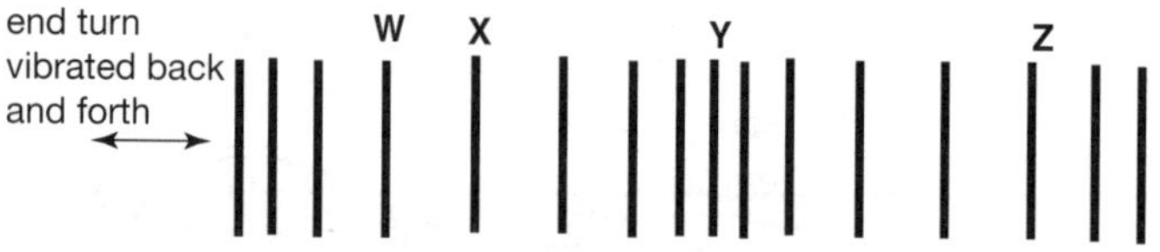

end turn vibrated back and forth
W X Y Z

Figure 2

Complete the statements to identify which of the turns, **W**, **X**, **Y** or **Z**, corresponds to the description given.

____________________ is displaced to the right of its undisturbed position.

____________________ is at the centre of a compression.

____________________ is at the centre of a rarefaction.

____________________ is displaced to the left of its undisturbed position. [4 marks]

Frequency and period

1. **Maths Skills**

A sound wave is produced by the vibrations of the cone of a loudspeaker. The period of the sound wave is 5.0 ms. Determine the frequency of vibration of the loudspeaker. Give the unit with your answer.

Frequency = ____________________

Unit ____________________ [3 marks]

Maths

The equation relating period (T) to frequency (f), $T = \frac{1}{f}$, is given on the Physics Equations Sheet. You are expected to be able to select and apply this equation – it may not be given in the question.

2. **a** State what is meant by the **wavelength** of a wave.

____ [1 mark]

Maths Skills

b **Figure 3** represents a snapshot of a transverse wave travelling on a rope.

Determine the wavelength of the wave.

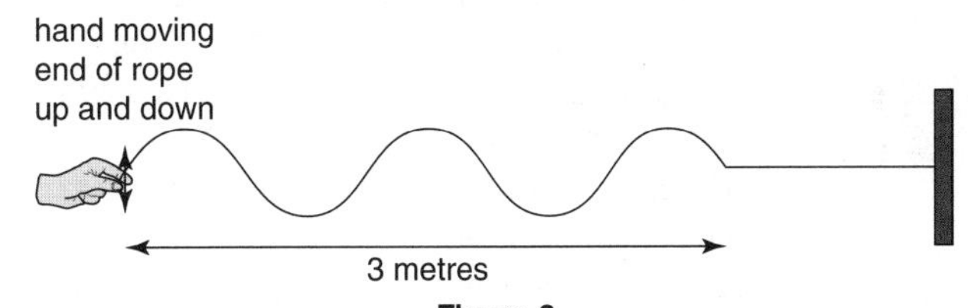

Figure 3

Wavelength = ____________________ m [1 mark]

3. **Maths Skills**

The lines in **Figure 4** represent wave crests of surface water waves produced by a vibrating beam. The lines represent wave crests travelling to the right on the water surface. The crest labelled **Y** was produced 4.0 s after the crest labelled **X**. Calculate the wavelength and period of the surface water wave.

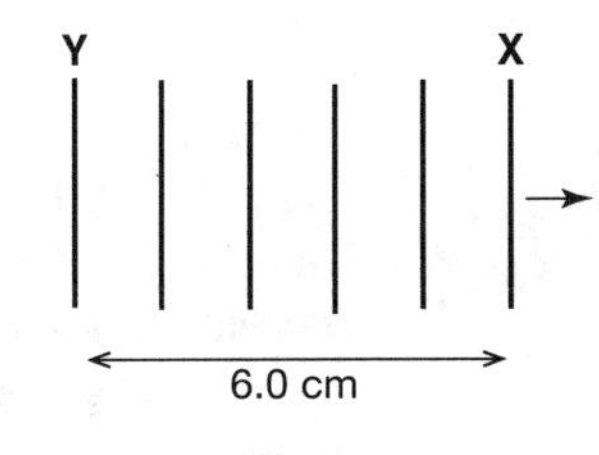

Figure 4

Wavelength = ____________________ cm

Period = ____________________ s [4 marks]

Wave speed

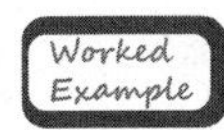

Surface water waves are produced in a ripple tank by a vibrating beam with a frequency of 10 Hz. The speed of the waves is measured as 30 cm/s. Calculate the distance between adjacent crests.

Distance between adjacent crests = $\lambda = \frac{v}{f}$

$= \frac{30 \text{ cm/s}}{10 \text{ Hz}} = 3.0 \text{ cm}$ Marks gained: [2 marks]

1.

Maths Skills

One end of a long slinky spring is moved back and forth at a frequency of 2.0 Hz to produce a longitudinal wave. A compression is observed to move 4.0 m in 2.5 s. Calculate the distance from the centre of one compression to the centre of the next compression.

Distance = ________________ m [3 marks]

Maths skills

The equation that links wave speed to frequency and wavelength is $v = f\lambda$. You need to learn this equation as it may not be given in the question.

2.

Maths Skills

In an experiment to measure the speed of sound in air, a student stands 80 m from a high wall and repeatedly claps two blocks of hard wood together. The student adjusts the clapping rate so that each clap coincides with the echo of the previous one. Another student then measures the time for 30 intervals between claps as 13.8 s. Calculate the speed of sound. Give your answer to two significant figures.

Speed = ________________ m/s [3 marks]

3.

Required practical

Two students are asked to test the equation $v = f\lambda$ for waves on a water surface. The waves are produced in a ripple tank using a vibrating beam driven at a low frequency (**Figure 5**). An LED strobe light is positioned above the tank and a large sheet of paper is placed under the tank. A shadow of the waves appears on the paper. The size of each shadow is greater than the actual size of the object that causes the shadow. One student places a 30 cm metal rule in the tank and measures the length of its shadow on the paper. The metal rule is then removed from the tank.

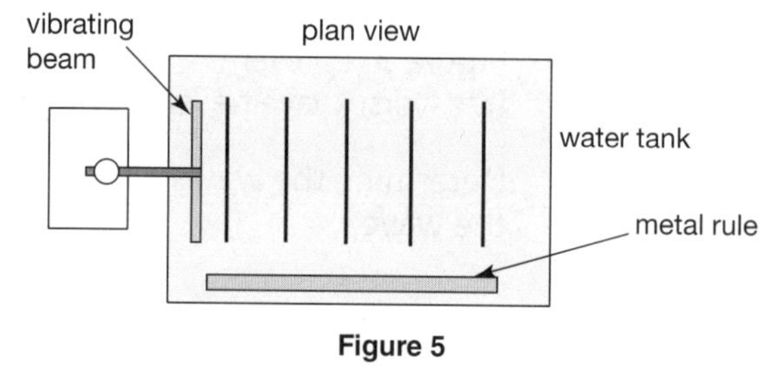

Figure 5

The student uses a stopwatch to measure the time for 20 vibrations of the beam. She then switches on the strobe light and adjusts the flash rate until the shadows of the waves on the paper appear stationary. Another student measures the distance across four wavelengths on the wave shadows using a metre rule on the paper. He then adjusts the LED strobe light so that it shines continuously. The image of the waves on the paper is now moving. He draws two marks on the paper, 50 cm apart. Using a stopwatch, the students measure the time taken for the image of one line of crests to travel between the two marks. The students repeat each of the time measurements. The students' time measurements are shown in **Table 1** and the distance measurements are shown in **Table 2**.

Measurement	First time reading (s)	Second time reading (s)
Time for 20 vibrations of the beam	9.4	9.6
Time for the image of a line of crests to travel 50 cm	2.4	2.0

Table 1

Measurement	Distance (cm)
Distance occupied by 4 wavelengths observed on the paper	42
Length of the image of the 30 cm metal rule	40

Table 2

Use the data in **Tables 1** and **2** to calculate the frequency, wavelength and speed of the **actual** water waves in the tank. Give your answers to two significant figures. Write down the equations that you use.

Frequency = ______________ Hz

Wavelength = ______________ cm

Speed = ______________ cm/s [7 marks]

Reflection and refraction of waves

1. **Figure 6** represents three processes, labelled **A**, **B** and **C**, that can occur when a ray of sunlight is incident on a pane of glass.

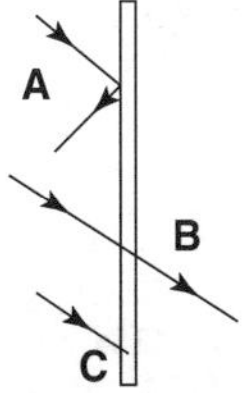

Figure 6

Name each process.

A: ______________

B: ______________

C: ______________

[3 marks]

2. Required practical

a A student is asked to use a plane mirror to demonstrate the law of reflection: *angle of incidence = angle of reflection*. Describe the experimental method that the student should follow.

____

____

____

____

____

[6 marks]

b A student obtains some values for angles of incidence (*i*) and reflection (*r*) at a plane mirror (**see the table**). Explain whether or not the data agrees with the law of reflection.

Angle *i* (degrees)	Angle *r* (degrees)
29	31
42	40

[2 marks]

3. Required practical

A student investigates what happens when a beam of light from a ray box is incident on the edge of a rectangular glass block (**Figure 7**). The diagram shows the normal at the point of incidence.

Add lines to represent the path of the resulting light rays that the student should expect to observe. [3 marks]

Figure 7

Sound waves

1. Higher Tier only

a A student puts his ear to the bench while a second student lightly taps further along the bench with a pencil. The sound appears louder and sharper to the first student than if he listens to the same sound through the air. State what this observation suggests about the transfer of a vibration from one molecule to the next molecule.

[1 mark]

Higher Tier only

b State the normal range of human hearing.

________ Hz to ________ Hz [2 marks]

Higher Tier only

c Suggest what happens to the energy carried by a sound wave when it enters a person's ear if the frequency is outside the range of human hearing.

____

____ [2 marks]

2. A tuning fork produces a sound of frequency of 520 Hz. Calculate the difference in the wavelength of the sound it produces when sounded indoors at 20 °C and then outdoors at 0 °C. Give your answer to two significant figures.

Higher Tier only

Speed of sound in air at 20 °C = 343 m/s

Speed of sound in air at 0 °C = 331 m/s.

Change in wavelength = ____________________ m [6 marks]

Ultrasound and echo sounding

1. **a** An echo sounder on board a ship has a frequency of 50 kHz. Calculate the depth of the ocean below the ship if the return pulse is detected after 1.92 s.

Maths Skills

Speed of sound in sea water = 1 480 m/s. Give your answer to three significant figures.

Higher Tier only

Depth = ____________________ m [3 marks]

Higher Tier only

b A higher frequency echo sounder, typically 200 kHz, is used in shallower water (up to 60 m) because it can detect much finer detail. Suggest a use for this higher frequency echo sounder.

____

____ [1 mark]

2. **a** Describe what happens when ultrasound waves reach a boundary between two different materials.

Higher Tier only

____

____

____ [2 marks]

Maths Skills

Higher Tier only

b Ultrasound waves are used in industry to detect cracks in materials. The metal block in **Figure 8** is 0.20 m thick (top to bottom). The ultrasound waves would be reflected back from both the crack and the lower surface of the metal block.

Use the data provided to explain how it is possible to distinguish between these two reflections.

Speed of sound in the metal = 6 100 m/s.

Figure 8

____

____

____

____

____ [3 marks]

Seismic waves

1.

Higher Tier only

a Seismic waves are produced by earthquakes. There are two types of seismic wave: P-waves and S-waves. State 'yes' or 'no' in the table to show the properties of each type of wave.

Wave	Transverse	Longitudinal	Travels though solids	Travels through liquids
P				
S				

[2 marks]

Higher Tier only

b Which type of seismic wave involves the ground moving perpendicular to the direction of energy transfer? Tick **one** box.

P-waves ☐ S-waves ☐ [1 mark]

Higher Tier only

c Which type of wave has the highest speed in a given material? Tick **one** box.

P-waves ☐ S-waves ☐ [1 mark]

2. Higher Tier only

a **Figure 9** represents the observation that S-waves pass through some parts of the Earth but not others.

What does this indicate about the Earth's core? Explain your answer.

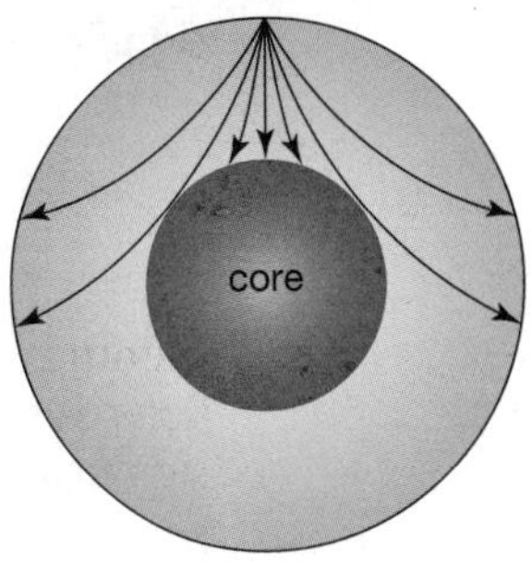

earthquake

core

Figure 9

____

____

____

____

______________________________ [3 marks]

b **Figure 10** represents the observation that P-waves undergo sudden changes of direction at certain locations in the Earth.

Give the name of this effect. State the cause of the effect.

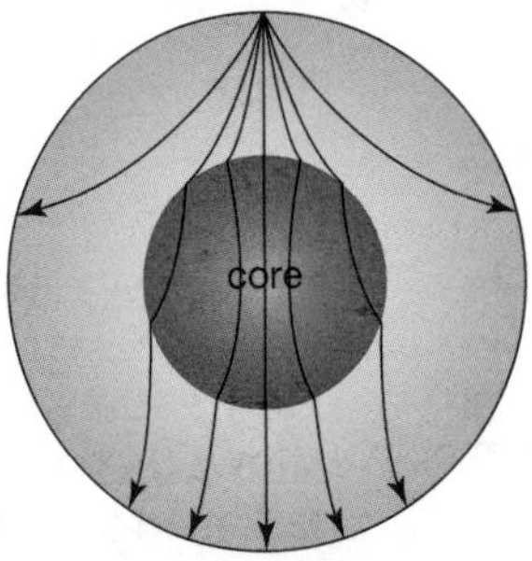

earthquake

core

Figure 10

____

____

____

______________________________ [2 marks]

The electromagnetic spectrum

1. Synoptic

Blue light has a wavelength of 4×10^{-7} m. Microwaves used in microwave cooking have a frequency around 2.5 GHz. Calculate by how many orders of magnitude the wavelengths of the microwaves and the blue light differ.

Speed of light = 3.0×10^{8} m/s.

Number of orders = ______________ [3 marks]

2. Electromagnetic waves travel at 3.0×10^8 m/s in a vacuum.

Maths Skills

a The closest star to Earth, not including the Sun, is Proxima Centauri. The distance from Proxima Centauri to Earth is approximately 4×10^{16} m. Calculate the time taken for the light from Proxima Centauri to travel to the Earth. Give your answer in years to one significant figure.

Time = ____________________ years [4 marks]

b Gamma rays of frequency 2.2×10^{26} Hz have been detected from sources beyond the Solar System. Calculate the wavelength of these gamma rays. Give your answer to two significant figures in standard form.

Wavelength = ____________________ m [3 marks]

Absorption, transmission, refraction and reflection of electromagnetic waves

1. Higher Tier only

Electromagnetic radiation from space is incident on the Earth's atmosphere. **Figure 11** shows that the fraction of the radiation that is absorbed by the atmosphere depends on its wavelength.

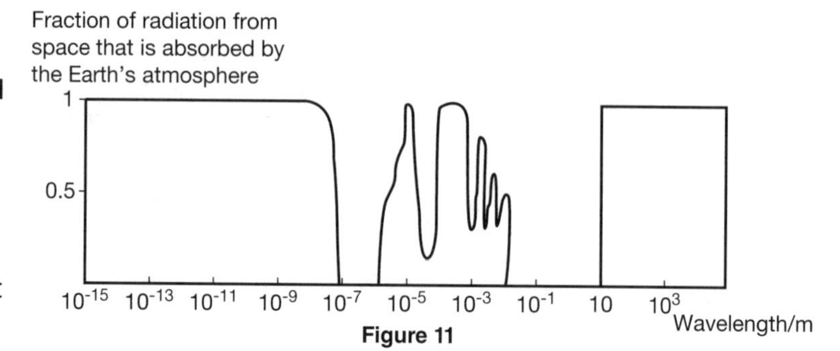

Figure 11

Which type of electromagnetic radiation is transmitted through the Earth's atmosphere with the lowest amount of absorption? Tick **one** box.

☐ Long wave radio waves ☐ Gamma rays ☐ Visible light [1 mark]

2. Higher Tier only

A laser beam is incident on a block of glass. In **Figure 12** the laser beam is represented by a series of wave fronts and the direction of energy transfer is shown by the arrow. The diagram is not to scale.

Add **three** wave fronts inside the block in **Figure 12** to show the refraction of the laser beam. Add an arrow to show the direction of energy transfer of the beam inside the glass. Note that the laser beam travels more slowly in the glass than in the air.

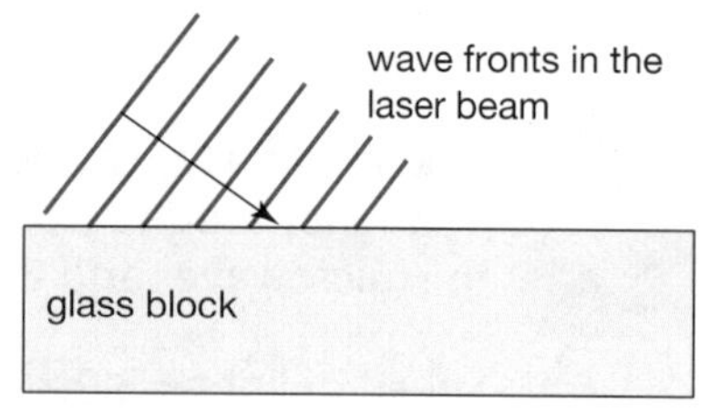

Figure 12

[3 marks]

3. **Figure 13** shows a ray of light incident on a section of double glazing. Use a ruler to draw the path taken by the light ray until it emerges from the second pane of glass.

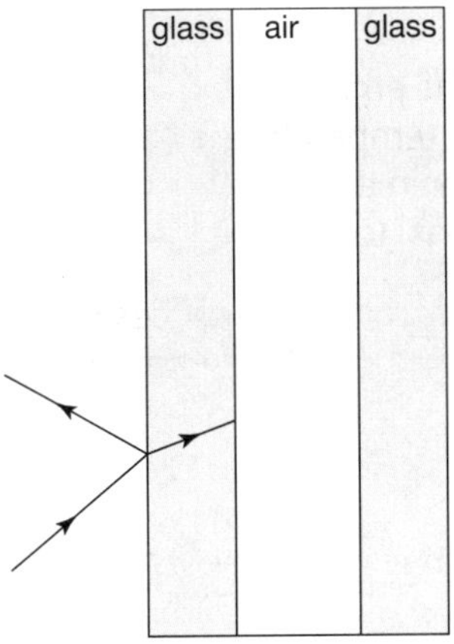

glass | air | glass

Figure 13

[4 marks]

4. **Higher Tier only**

White light is made up of several different colours. A student investigates what happens to white light when it is incident on a glass prism. **Figure 14** shows what happens to the red and blue light.

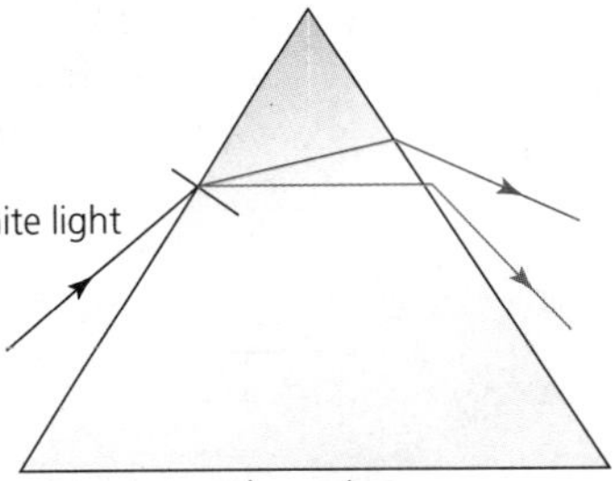

white light

glass prism

Figure 14

a State what happens to the speed of the light on entering the glass.

______________________________ [1 mark]

b State which has the shorter wavelength, red or blue light.

______________ [1 mark]

c Suggest what can be concluded about the relationship between the wavelength of the light and the amount of refraction that occurs on passing through the glass prism.

____

____

______________________________ [1 mark]

Emission and absorption of infrared radiation

1. Required Practical

A student uses the apparatus in **Figure 15** to investigate how the emission of infrared radiation from a blackened metal surface varies with temperature. A thermometer is used to measure the temperature of the water in the container as it cools. The voltmeter reading is directly proportional to the intensity of the infrared radiation.

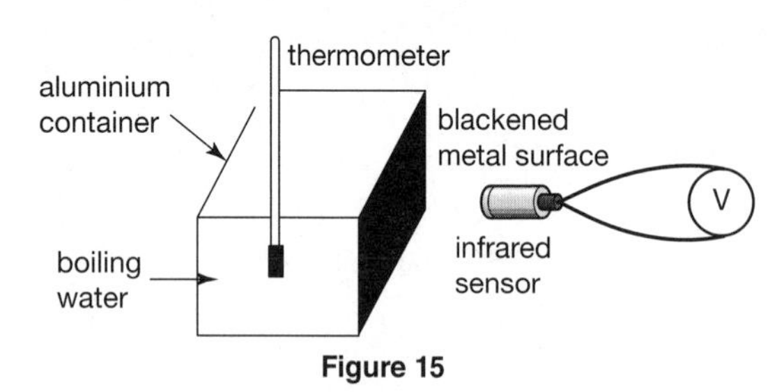

Figure 15

a The student uses the apparatus to investigate the amount of infrared radiation emitted from the different sides of the aluminium container. State a control variable for this experiment.

1 ______________________________

2 ______________________________ [2 marks]

b **Table 5** shows the student's measurements.

Surface	Polished aluminium	Dull aluminium	Shiny black	Dull/matt black
Voltage (mV)	6	46	60	67

Table 5

State what can be concluded from this experiment.

____

____

____

______________________________ [2 marks]

Uses and hazards of the electromagnetic spectrum

1. Synoptic Higher Tier only

Nuclei of the radioisotope technetium-99m emit only gamma radiation. Explain why this property of technetium-99m makes it suitable for use as a tracer for producing images of internal organs.

[3 marks]

2. Higher Tier only

a X-rays are used for producing images of bones to check for fractures or breaks. Explain what makes X-rays suitable for this application.

[2 marks]

b **Table 6** gives the radiation dose for two medical procedures that use X-rays.

Procedure	Radiation dose (mSv)
Chest X-ray	0.1
Full body CT scan	10

Table 6

State why it is necessary to know the radiation dose received by a patient when undergoing these procedures.

[1 mark]

Radio waves

1. Radio waves are used for transmitting television programs.

Higher Tier only

a Describe the basic principle of the production of radio waves.

____

____ [2 marks]

b **Figure 16** represents a typical aerial that is attached to the roof of a house to receive television signals transmitted by an Earth-based transmitter.

Describe how these radio waves are received.

Figure 16

____

____

____

____ [3 marks]

2.

Higher Tier only

a Microwaves are used for transmitting information between satellites and the Earth. Explain why microwaves are more suitable than radio waves for transmitting information in this way.

____

____

____ [3 marks]

Higher Tier only

b 'Long wave' radio waves are using for transmission of information between continents. 'Short wave' radio waves cannot be used for this purpose. Describe how the properties of long wave radio waves makes them suitable for this purpose.

____

____ [2 marks]

Colour

1. **a** **Figures 17a** and **17b** represent two types of reflection.

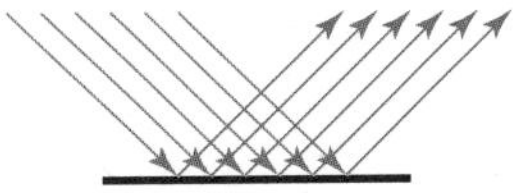

Give the names of these types of reflection shown in **Figures 17a** and **17b**.

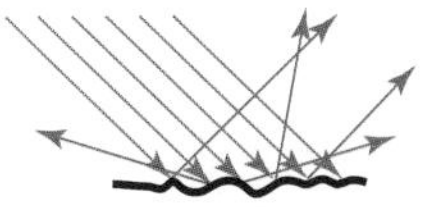

Figure 17a

Figure 17b

Figure 17a: ____________ Figure 17b: ____________ [2 marks]

b State what type of reflection is occurring when you see an image of yourself in a shop window.

____________ [1 mark]

2. **a** A particular opaque object appears green when illuminated with white light. Explain what happens to the white light that is incident on the opaque object.

____

____

____

____

____________ [3 marks]

b State how the object referred to in part (a) would appear if illuminated only by red light.

____________ [1 mark]

3. **a** Compare the use of the terms **transparent** and **translucent** to describe an object.

____

____

____

____________ [2 marks]

b A coloured filter is made from transparent dyed glass or plastic. Explain what is observed if you look at white light from a bulb through a red filter.

___

___

___ [3 marks]

Lenses

1. **a** State the type of lens shown in **Figure 18**.

Lens type: ___

[1 mark]

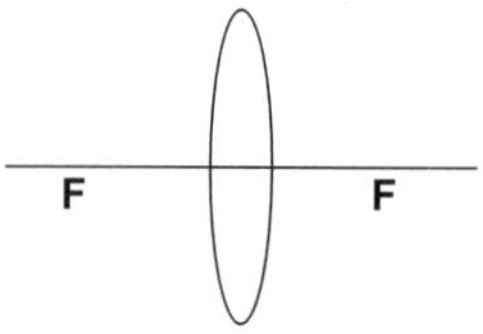

Figure 18

Maths

The equation for the magnification by a lens:

$$\text{magnification} = \frac{\text{image height}}{\text{object height}}$$

is on the Physics Equations Sheet. You are expected to be able to select and apply this equation – it may not be given in the question.

b The letter **F** in **Figure 18** represents the principal focus of the lens. Describe what is meant by the **principal focus** of a lens.

___

___ [1 mark]

c Describe what is meant by the **focal length** of a lens.

___ [1 mark]

d The lens inside a human eye produces an image on the retina. The lens has a focal length of about 2 cm. Complete the ray diagram in **Figure 19** to show how the lens forms an image of a distant object.

[3 marks]

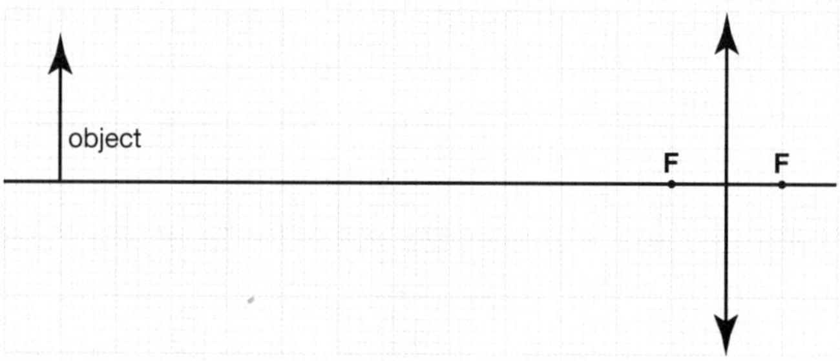

Figure 19

e Give two properties of the image formed.

___ [2 marks]

f Estimate the magnification of the image.

Magnification = ____________ [1 mark]

A perfect black body

1. Synoptic

a When the power supply unit in **Figure 20** is switched on the filament appears red.

Explain why.

Figure 20

____

____

____

____

____ [3 marks]

Synoptic

b The output potential difference of the power supply unit is increased. The light emitted from the filament is brighter and appears yellow. Explain the changes to the light emitted from the filament.

____

____

____

____ [4 marks]

2. a The theoretical concept of a perfect black body is useful to scientists. Explain what would happen to the light incident on a perfect black body.

____

____

____ [3 marks]

b A perfect black body would also be a perfect emitter of electromagnetic radiation. Explain what this means.

____

____

____ [2 marks]

Temperature of the Earth

1. The surface temperature of the Sun is about 5 500 °C. Most of its energy is radiated as visible light and short wavelength infrared. The average surface temperature of the Earth is 14 °C and most of the energy radiated from the Earth is longer wavelength infrared.

The average intensity of the radiation from the Sun reaching the top of the Earth's atmosphere is 342 W/m^2. **Figure 21** shows data for the different processes by which energy enters, leaves or is stored in the Earth's climate system.

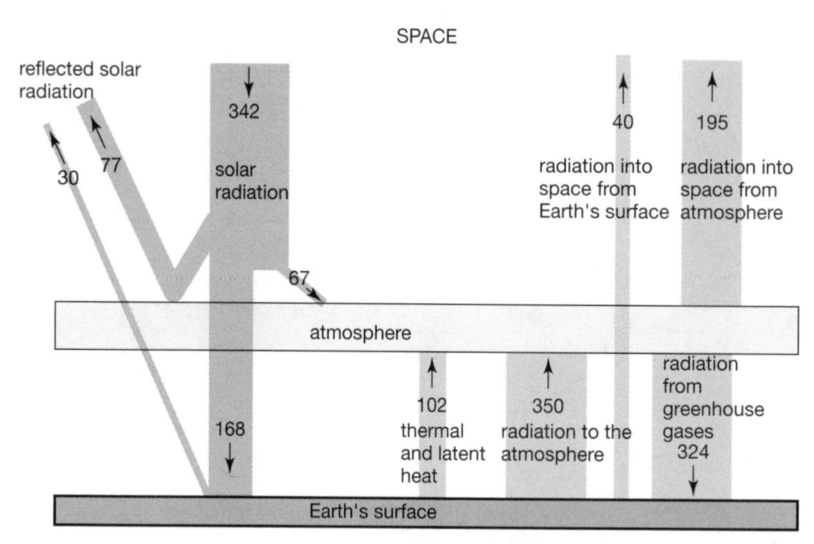

Figure 21

a Identify the processes in **Figure 21** that transfer energy **into** the Earth's surface. Give their total in W/m^2.

____

____

Total ____________ W/m^2 [2 marks]

b Identify the processes that transfer energy **away** from the Earth's surface. Give their total in W/m^2.

____

____

____

____

____

____

Total ________________ W/m^2 [4 marks]

c State what this data suggests about the average temperature of the Earth's surface.

____

____ [1 mark]

Magnets and magnetic forces

1. **a** Two bar magnets are placed on a table. The two north poles are 3 cm apart (**Figure 1**).

N N

Figure 1

Describe the force between the two north poles. Repulsive non-contact

The force between the two north poles is a repel force

[2 marks]

b The force between the two north poles in **Figure 1** is 65 mN. One of the magnets is rotated so that its south pole faces the north pole of the other magnet. The poles are placed 5 cm apart. What is the force between the magnets? Tick **one** box. [1 mark]

- [ ] 35 mN repulsive
- [x] 65 mN attractive X
- [ ] 35 mN attractive
- [ ] 70 mN attractive

2. **a** A student strokes a steel bar with the north pole of a known magnet (**Figure 2**).

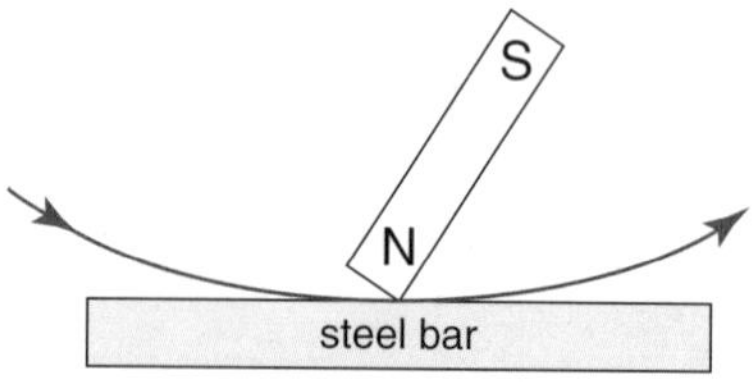

S

N

steel bar

Figure 2

Describe how the student could test the steel bar to determine if it had become a permanent magnet.

If one end is repelled by a pole of a known magnet, it has become permanent

A Student Could test the steel bar to determine if it had become a permanent magnet is to see if it repels.

[1 mark]

b The student now hangs two iron nails from the ends of the magnet (**Figure 3**).

N S

Figure 3

Explain why the nails lean towards each other.

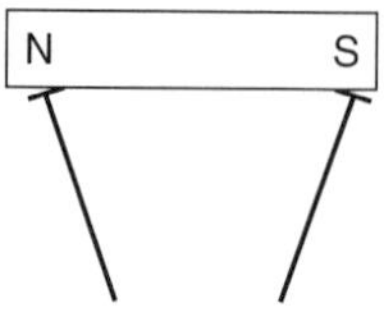

The nails lean towards each other due to the Magnetic force pushing towards the center

Each nail becomes an induced magnet because they are in the magnetic field of the bar magnet. The ends of the nails become opposite poles with a force of attraction between them.

[4 marks]

Magnetic fields

1. **a** Draw lines on **Figure 4** to represent the strength and direction of the magnetic field of the bar magnet. [4 marks]

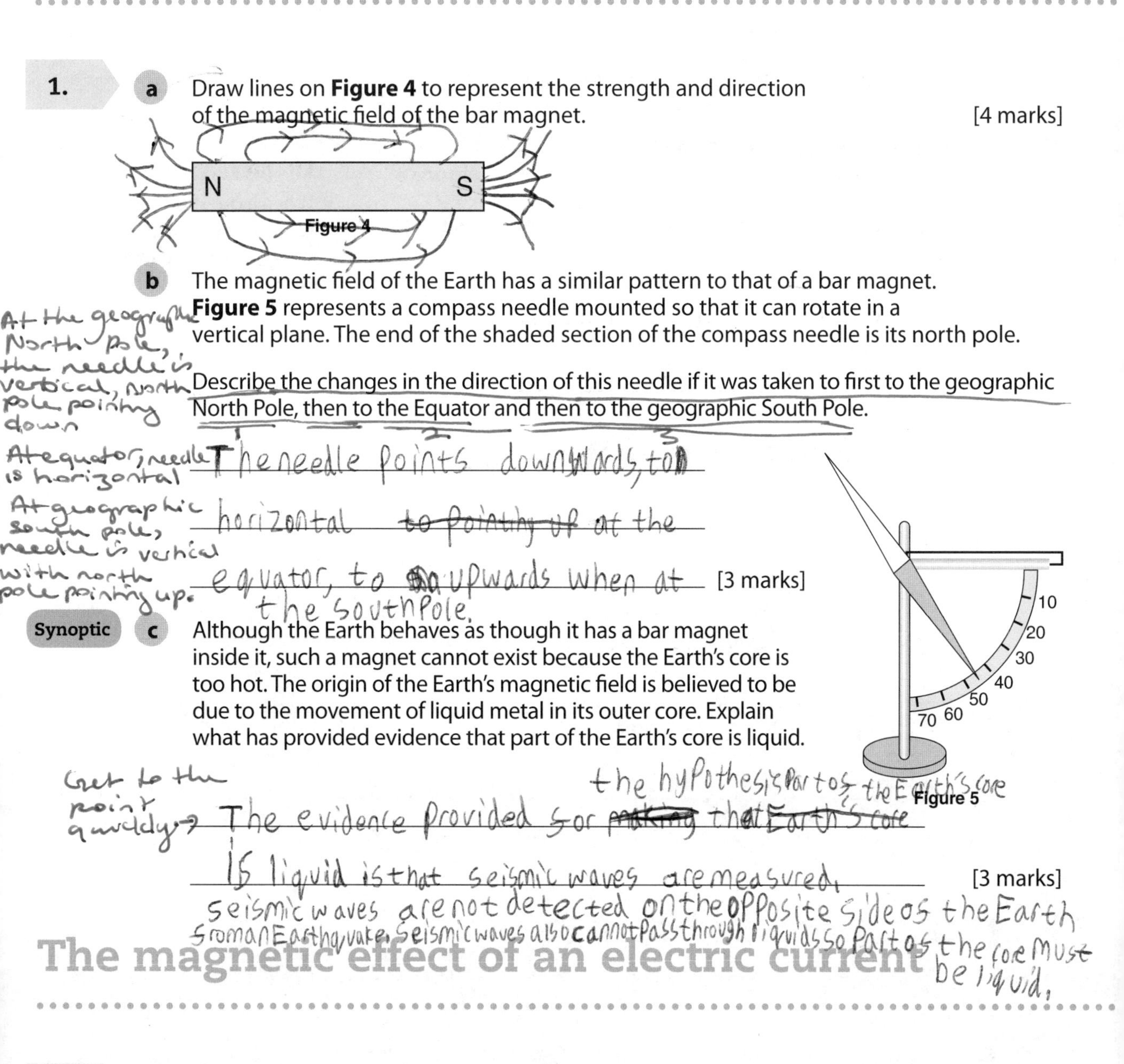

Figure 4

b The magnetic field of the Earth has a similar pattern to that of a bar magnet. **Figure 5** represents a compass needle mounted so that it can rotate in a vertical plane. The end of the shaded section of the compass needle is its north pole.

Describe the changes in the direction of this needle if it was taken to first to the geographic North Pole, then to the Equator and then to the geographic South Pole.

[3 marks]

Synoptic **c** Although the Earth behaves as though it has a bar magnet inside it, such a magnet cannot exist because the Earth's core is too hot. The origin of the Earth's magnetic field is believed to be due to the movement of liquid metal in its outer core. Explain what has provided evidence that part of the Earth's core is liquid.

Figure 5

[3 marks]

The magnetic effect of an electric current

1. **a** A plotting compass can be placed at different positions on the card in **Figure 6a** to show the magnetic field created by the wire when a current passes through the wire.

On **Figure 6b**, draw the magnetic field produced by the wire. Show the field direction clearly. [3 marks]

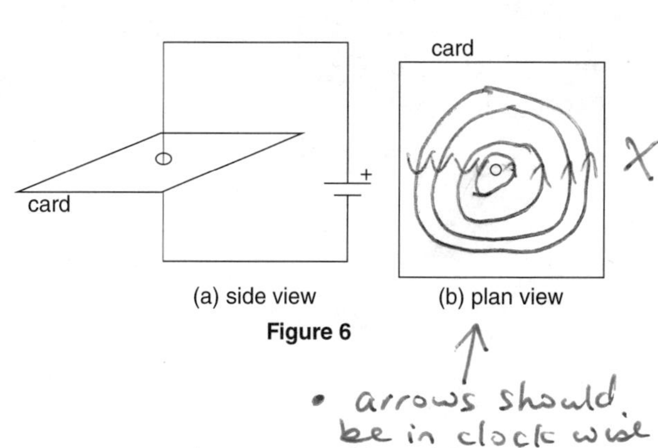

Figure 6

b The wire is now wrapped around a cardboard tube to create a solenoid (**Figure 7**).

Label the current direction in the solenoid, and the north and south poles of the solenoid's magnetic field when the switch is closed. [3 marks]

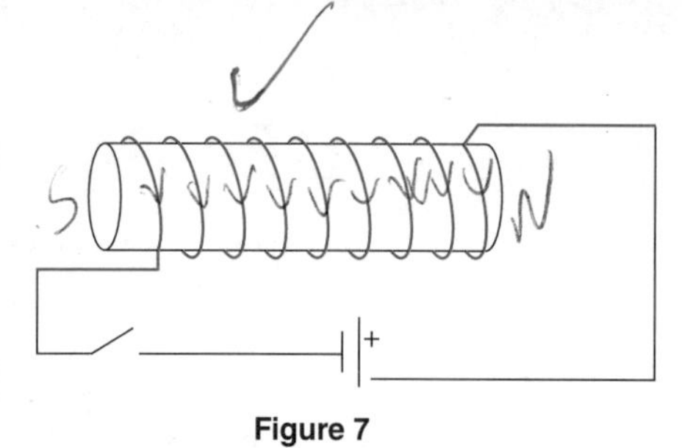

Figure 7

2. An electromagnetic switch can be used to switch the current on in another circuit such as a motor circuit (**M** in **Figure 8**).

Explain what happens from the moment that the switch (**S**) is closed.

When the switch is closed the current flows in the solenoid. The solenoids current creates a magnetic field. The magnetic field of the solenoid induces magnetism in the iron strips. The end of the iron strips in the middle of the solenoid have opposite angles. The iron strips are attracted together and make contact. The motor circuit is now complete. Current flows in the motor causing it to start. [4 marks]

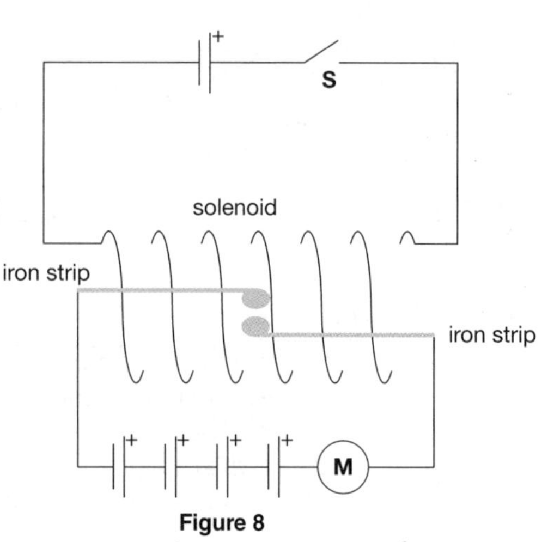

Figure 8

Literacy

When a question asks you to explain how a device works, make sure that each step is logically linked to the previous one.

Fleming's left-hand rule

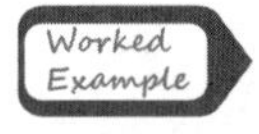

Higher Tier only

Calculate the magnetic force on a 5.0 cm wire carrying a current of 2.0 A in a magnetic field of magnetic flux density 0.25 T.

Force = $F = BIL$

$= 0.25 \times 2.0 \times 0.05$

$= 2.5 \times 10^{-2}$ N

marks gained: [0 marks]

Maths

The equation linking the force on a conductor carrying current, F, to magnetic flux density, B, current, I, and length, l, is $F = BIL$. You need to be able to select and apply this equation which is given on the Physics Equations Sheet.

1. **a** A piece of insulated copper wire is placed in a magnetic field (**Figure 9**).

Higher Tier only

Determine the direction of the force on the wire when the switch is closed.

Direction: ______________________ [1 mark]

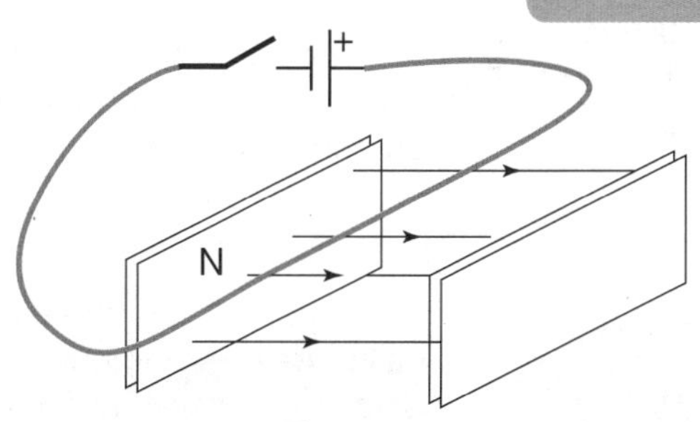

Figure 9

Maths Skills

Higher Tier only

b The current through the wire is 3.8 A. The magnetic flux density of the field is 300 mT. The length of wire within the field is 5.0 cm. Calculate the magnetic force on the wire. Give your answer in mN.

Force = ______________________ mN [3 marks]

2. **Figure 10** represents a simple device for measuring magnetic flux density. The copper wire frame is pivoted at the centre of its longer sides and balances horizontally when the switch is open. The length of copper wire between the poles of the horseshoe magnet is 2.0 cm long.

Higher Tier only

Synoptic

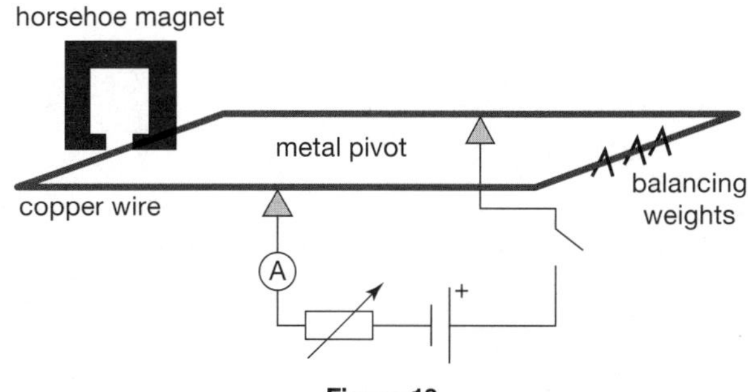

Figure 10

Three small weights, each of mass 3.0×10^{-3} kg, are hooked over the right-hand side of the copper wire frame. The switch is closed and the variable resistor adjusted until the copper frame is again balanced horizontally. At this point the ammeter reads 4.1 A. Calculate the magnetic flux density between the poles of the horseshoe magnet. Write down the equations that you use. Give your answer to two significant figures.

Gravitational field strength = 9.8 N/kg

Magnetic flux density = ______________________ T [5 marks]

Electric motors

1. **a** **Figure 11** represents a d.c. electric motor showing a one turn of a coil. Determine the directions of the forces on the left-and right-hand sides of the coil and the resulting direction of rotation of the coil.

Higher Tier only

Left side force direction: ______________

Right side force direction:

____

Rotation: ______________ [3 marks]

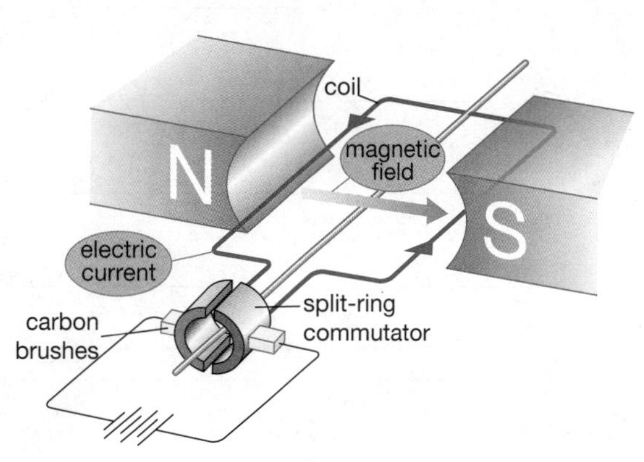

Figure 11

Higher Tier only

b The current in the coil is 0.20 A. The length of the section of the coil within the field is 4.0 cm and the magnetic flux density of the field is 0.30 T. Calculate the size of the force on one side of the coil.

Force = ____________________ [2 marks]

Higher Tier only

Synoptic

c The distance from the left side of the coil to the central axis of the motor is 1.8 cm. Calculate the moment created by the magnetic force on the left-hand side of the coil. Give your answer in N m to two significant figures.

Moment = ____________________ N m [3 marks]

Higher Tier only

d Calculate the total moment acting on the coil.

Total moment = ____________________ [2 marks]

Loudspeakers

1. An alternating potential difference is applied to the coil of the loudspeaker in **Figure 12**.

Higher Tier only

Explain how the loudspeaker produces a sound.

S
N
S
paper cone
alternating potential difference

Figure 12

____

____

____

____

____

____

____

____

____

____ [6 marks]

Induced potential

1. Higher Tier only

Figure 13 shows a wire placed between two magnets. As the wire in **Figure 13** is raised upwards, an electric current is induced in the wire.

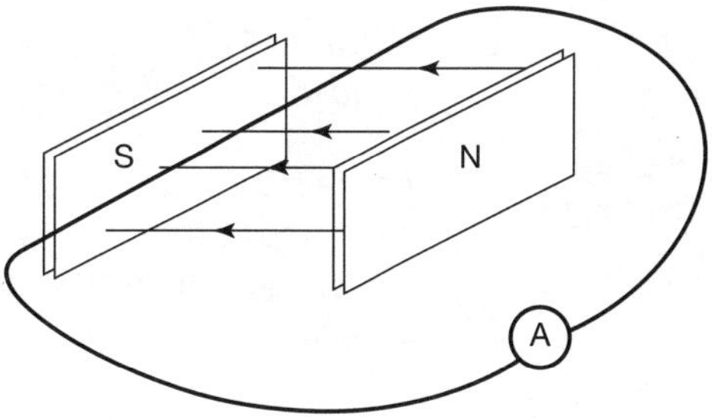

S N A

Figure 13

a Describe **two** changes that would each result in a larger current being induced in the wire.

[2 marks]

Higher Tier only

b Describe **two** changes that would each cause the current direction to be reversed.

[2 marks]

2. Higher Tier only

As the magnet in **Figure 14** is moved downwards towards the coil, the top end of the coil behaves like the north pole of a magnet.

Explain what is occurring.

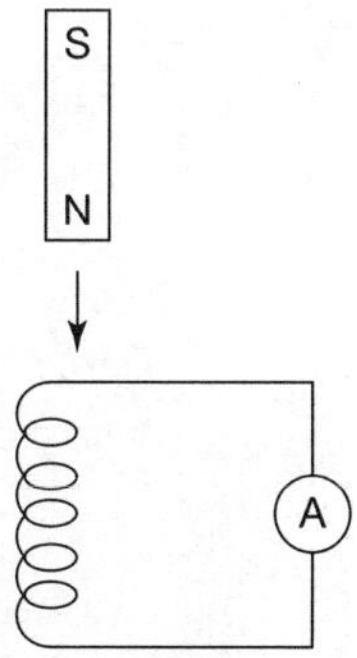

S N A

Figure 14

[4 marks]

Uses of the generator effect

1. **a** **Higher Tier only**

Figure 15 represents a simple hand-driven a.c. generator.

As the handle is rotated, describe what occurs that results in a potential difference being generated.

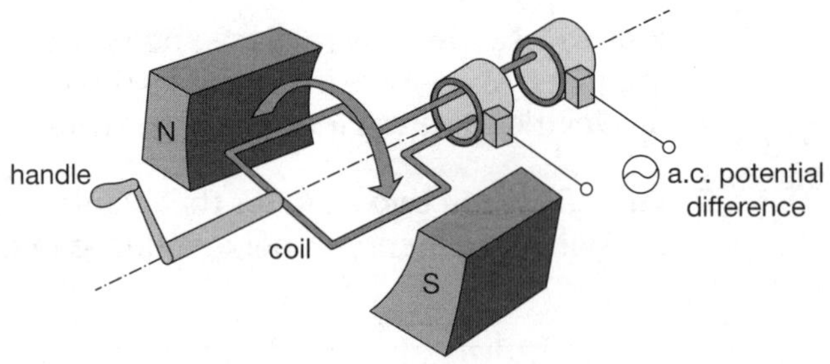

handle
N
coil
S
a.c. potential difference

Figure 15

____

____

____

____

____ [2 marks]

Command words

Although **sketch** means to draw approximately, you should still take care with the drawing and you still need to use the data given in the question when drawing the graph.

b **Higher Tier only**

On the axes in **Figure 16**, sketch the output potential difference produced when the coil is rotated once in 1 second.

[2 marks]

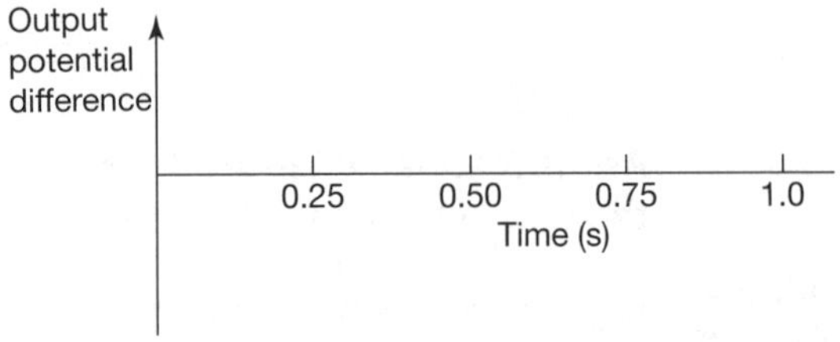

Output potential difference
0.25 0.50 0.75 1.0
Time (s)

Figure 16

c **Higher Tier only**

State whether the maximum output voltage occurs when the coil is horizontal or vertical.

____ [1 mark]

2. **Higher Tier only**

The variation in the output potential difference of a d.c. generator as time passes is shown in **Figure 17**.

Output potential difference
0.01 0.02 0.03 0.04
Time (s)

Figure 17

Determine the number of revolutions the coil of the generator makes in 1 second.

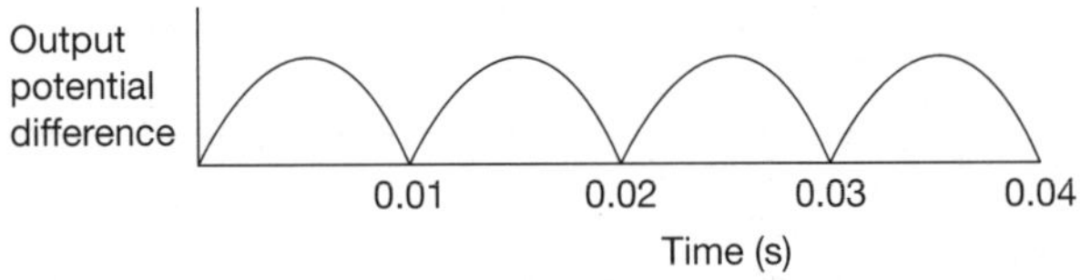

Number = ______________ revolutions per second [2 marks]

Microphones

1. A moving coil microphone is shown in **Figure 18**.

Higher Tier only

a Explain how the output potential difference is generated.

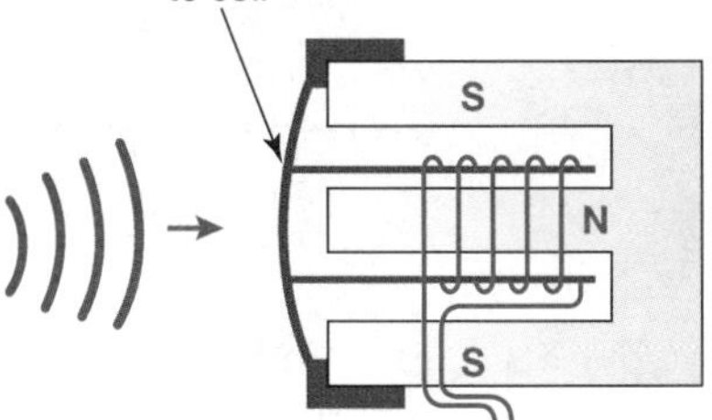

[3 marks]

Figure 18

b Describe **one** similarity between the incoming sound wave and the output potential difference.

[1 mark]

Transformers

1. The potential difference needed to operate a particular laptop is 15.0 V. The laptop's battery is charged by connecting it to the mains via a transformer.

Higher Tier only

a Explain why the type of transformer shown in **Figure 19** is suitable for charging the laptop's battery from the 230 V mains supply.

Maths

You need to be able to select and apply the two transformer equations: $\frac{V_p}{V_s} = \frac{n_p}{n_s}$ and $V_s I_s = V_p I_p$ from the Physics Equations Sheet.

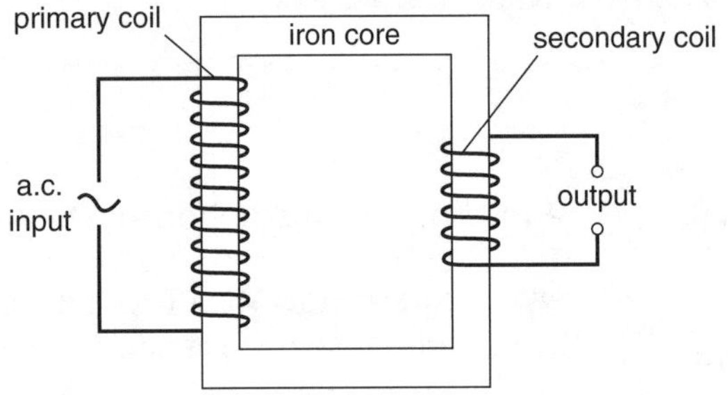

Figure 19

[2 marks]

b Explain how an alternating voltage output is created at the output of the secondary coil when the primary coil is connected to the mains.

[3 marks]

Maths Skills

c The transformer used for charging the 15.0 V laptop battery from the mains has a 500 turn primary coil. Calculate the number of turns on the transformer's secondary coil. Give your answer to the nearest whole number. Use the correct equation from the Physics Equations Sheet.

Number of turns = ____________ [4 marks]

Maths Skills

d While the laptop battery is being charged the current in the transformer's secondary coil is 8.16 A. Calculate the current in the transformer's primary coil. Assume that the transformer is 100% efficient. Give your answer to three significant figures. Use the correct equation from the Physics Equations Sheet.

Primary current = ____________ A [4 marks]

Our Solar System

1. Eight planets and at least three dwarf planets orbit the Sun. What force keeps a planet in orbit around the Sun? Describe the nature of this force.

___

___

___ [3 marks]

Maths

Orders of magnitude are usually written as powers of 10 and can be used to make approximate comparisons. If two values differ by two orders of magnitude, then one value is approximately 10^2 (or 100) times bigger than the other.

2. Maths Skills

An estimate of the radius of our Solar System is 10^{12} m. An estimate of the radius of our Milky Way galaxy is 10^{17} m. Determine by how many orders of magnitude the Milky Way and the Solar System radii differ.

Orders of magnitude = ___ [1 mark]

The life cycle of a star

1. Describe how atoms of all the naturally occurring elements up to and including iron are created in a star.

___

___

___

___

___

___

___ [4 marks]

2. **Figure 1** shows a graph of the time a star spends in its main sequence stage versus its mass, for stars up to 2 solar masses.

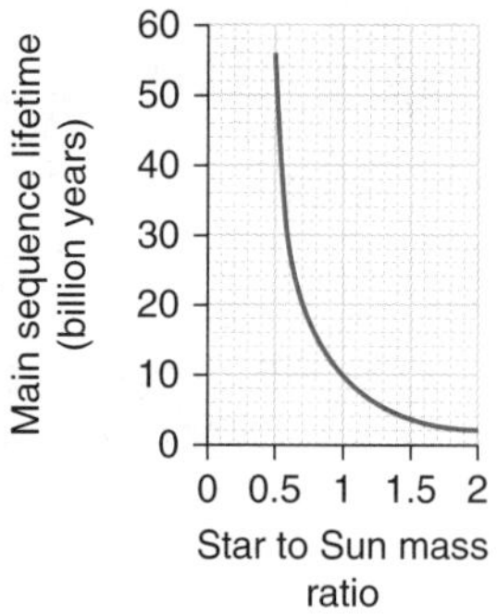

a Describe the relationship between the mass of a star and the time it spends as a main sequence star as shown by **Figure 1**.

____

______________________________________ [2 marks]

Figure 1

Maths Skills

b Use the graph to predict the main sequence lifetime of a star that has a mass 50% greater than the mass of the Sun.

Lifetime = ______________ years [1 mark]

3. A star of about the same mass as the Sun spends about 10 billion years as a main sequence star. Describe the stages in the life of such a star after it ceases to be a main sequence star.

____

____

____

____

______________________________________ [6 marks]

Orbital motion, natural and artificial satellites

1. Higher Tier only

a The eight planets of the solar system orbit the Sun in orbits that are almost circular. Explain how the velocity of a planet can be constantly changing even though its orbital speed remains the same.

____

______________________________________ [2 marks]

Maths Skills

b The circumference of the Earth's orbit around the Sun is 9.40×10^{11} m. The Earth takes 365 days to orbit the Sun. Calculate the Earth's orbital speed. Give your answer to three significant figures in standard form.

Orbital speed = ______________ m/s [4 marks]

2. **Higher Tier only**

a An artificial satellite is carried by a rocket to approximately the correct orbit and released. Small rockets on board the satellite are then fired to adjust its speed.

Explain what would happen to the satellite if its speed was not the correct value.

___

___

___ [2 marks]

Higher Tier only

b The table gives information about two artificial satellites that are either currently or have previously been in orbit around the Earth.

Satellite	Height of orbit above the Earth/km	Type of orbit	Time for one orbit of the Earth
MetOp-B	810	polar	100 minutes
Inmarsat-4	35 800	geostationary	24 hours

The polar orbit of MetOp-B enables it to observe the whole of the Earth. It is used to forecast the weather. The Inmarsat-4 satellite is used to deliver high-speed mobile broadband communications. Its geostationary orbit keeps it 'fixed' above a specific area of the Earth. Suggest why the orbit chosen for MetOp-B is more suitable for a weather satellite than the orbit of Inmarsat-4.

___

___

___

___ [2 marks]

Red-shift

1. a An astronomer used a telescope to observe the light from a distant galaxy. He observed that the light was red-shifted. Explain what **red-shift** means with respect to the motion of this distant galaxy.

___

___ [2 marks]

b The astronomer was able to measure the speed of recession of the distant galaxy and its distance from the Earth. He made speed and distance measurements for two other distant galaxies. The data is recorded in the table.

Speed (km/s)	Distance from Earth (km)
10 000	4.3×10^{21}
14 000	6.8×10^{21}
22 000	1.1×10^{22}

What does the data in the table suggest about the relationship between recession speed and distance for distant galaxies?

____

______________________________ [1 mark]

2. **Figure 2** shows a best-fit straight line graph of speed versus distance for distant galaxies. The distance axis is in megaparsecs (Mpc), which is a unit used for astronomical distances.

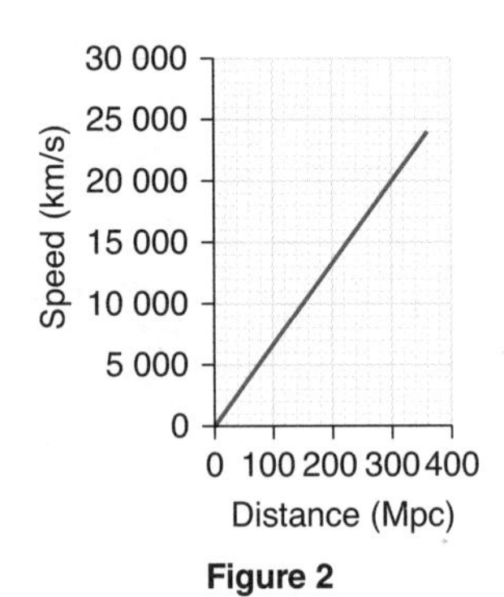

Figure 2

Maths Skills

a Use **Figure 2** to predict the speed of a galaxy at a distance of 400 Mpc from the Earth.

Speed = ______________ km/s [1 mark]

Maths Skills

b An astronomer uses the red-shift of the light from a distant galaxy to determine its recession speed as 10 000 km/s. The astronomer is unable to measure the galaxy's distance from Earth. Use **Figure 2** to predict the distance to this galaxy.

Distance = ______________ Mpc [1 mark]

3. Explain how observation of the red-shift of the light from distant galaxies supports the Big Bang model of the Universe.

____

____

____

____

____

______________________________ [4 marks]

Dark matter and dark energy

1. Describe what is meant by **dark matter**.

______

______

______ [2 marks]

2. **a** The Hubble Space Telescope can make observations of astronomical bodies at huge distances from the Earth. Explain why observing a very distant astronomical body is effectively looking back in time.

______

______

______ [2 marks]

b The Hubble Space Telescope has been used to measure the red-shift and distance of supernovae in very distant galaxies. The red-shift and distance measurements of the supernovae suggest that the expansion of the Universe is faster now than it was in the past.

Describe the concept introduced by scientists to explain the accelerated expansion of the Universe.

______

______

______ [2 marks]

Section 1: Energy changes in a system

Energy stores and systems

1. a Energy transfers from the tea's/hot liquid's thermal energy store [1 mark] to the thermal energy store of the surroundings. [1 mark]
   b Energy transfers from the chemical energy store of the fuel [1 mark] to the kinetic energy store of the car. [1 mark]
   c Energy transfers from the elastic potential energy store [1 mark] of the catapult to the kinetic energy store of the stone [1 mark]. On striking the wall, energy transfers from the kinetic energy store [1 mark] of the stone to the thermal energy store of the stone/wall. [3 marks]

Calculating energy changes

1. a $E_k = \frac{1}{2} \times 850 \times 12^2$ [1 mark]; 6.1×10^4 J [1 mark]; correct to two significant figures as shown. [1 mark]
   b $v = \sqrt{\frac{2E_k}{m}} = \sqrt{\frac{2 \times 12.2 \times 10^4}{850}}$ [1 mark] 17 m/s [1 mark]; correct to two significant figures as shown. [1 mark]
2. a $E_p = 0.057 \times 9.8 \times 1.8$ [1 mark]; 1.0 J [1 mark]; correct to two significant figures as shown. [1 mark]
   b $E_k = 0.8 \times 1.0 = 0.80$ J [1 mark]; $v = \sqrt{\frac{2E_k}{m}}$ [1 mark for rearranged equation]; $\sqrt{\frac{2 \times 0.80}{0.057}}$ [1 mark]; 5.3 m/s [1 mark]; correct to two significant figures as shown. [1 mark]
3. $\frac{1}{2} \times 50 \times 0.2^2$ [1 mark] 1.0 J (accept 1) [1 mark]
4. a $E_p = mgh = 0.05 \times 9.8 \times 2$ [1 mark]; 0.98 J [1 mark]
   b $0.98 + (0.82 \times 1.0)$ [1 mark]; 1.8 J [1 mark]

Calculating energy changes when a system is heated

1. a $c = \frac{\Delta E}{m\Delta\theta}$ [1 mark for rearranging] $\frac{21260}{41.3-18.4}$ [1 mark]; 928 J/kg °C [1 mark] (no significant figure penalty).
   b To reduce the energy transferred to the thermal energy store of the surrounding air. [1 mark]
2. a Thermal energy = $1.25 \times 4200 \times (100 - 20)$ [1 mark]; 420 000 J. [1 mark];
   b Energy to body of the kettle = $0.8 \times 500 \times (100 - 20)$ [1 mark] 32 000 J. [1 mark];
   Total energy supplied 420 000 + 32 000 [1 mark] 452 000 J. [1 mark]
   c A minimum [1 mark]; Some of the energy supplied will be transferred to the thermal energy store of the surroundings. [1 mark]
3. a $\Delta E = m\,c\,\Delta\theta = 0.1 \times 4200 \times (23.5 - 20.0)$ [1 mark]; 1470 J [1 mark]
   b $\Delta c = \frac{\Delta E}{m\Delta\theta}$ [1 mark] $= \frac{1470}{0.05 \times (100.0 - 23.5)}$ [1 mark]; 384 J/kg °C [1 mark] (no significant figure penalty).

Work and power

1. a Work = $5200 \times 50 = 260\,000$ J. [1 mark]
   b Energy is transferred from the car's store of kinetic energy [1 mark] to the brakes' store of thermal energy. [1 mark]
2. Energy transferred to the load = $mgh = 600 \times 9.8 \times 25$ [1 mark]; 147 000 J. [1 mark]
   Useful output power = $\frac{147000}{49}$ [1 mark]; 3000 W. [1 mark]

Conservation of energy

1. a The transfer of energy to a store from which it cannot be retrieved. [1 mark]
   b Kinetic energy is transferred from the bicycle to thermal energy store of the surroundings. [1 mark]
2. a Work = $150 \times 20 \times 60$ [1 mark]; 180 000 J. [1 mark]
   b Energy from food = $4 \times 180\,000 = 720\,000$ J. [1 mark]

Ways of reducing unwanted energy transfers

1. a Any suitable insulator [1 mark] (e.g. expanded polystyrene, fibre glass, straw).
   b Dependent: temperature of the water (in the smaller beaker) [1 mark]; Independent: time. [1 mark]
   c Volume of hot water [1 mark]; initial water temperature [1 mark]; thickness of insulator. [1 mark]

d The volume of hot water should be measured in a measuring cylinder (before being transferred to the small beaker) [1 mark]; The stopclock should only be started when the hot water is at a chosen temperature [1 mark]; Sufficient insulator must be used each time so as to fill the gap between the two beakers. [1 mark]

e Plot all the data sets on a graph of temperature on the *y*-axis versus time on the *x*-axis [1 mark]; The best insulator will have the highest temperature at any given value of time (before room temperature is reached). [1 mark]

Efficiency

1. Energy = $800 \times \frac{15}{100}$ [1 mark]; 120 MJ. [1 mark]

2. a Energy = mgh [1 mark]; $0.5 \times 9.8 \times 0.8$ [1 mark]; 3.9 J. [1 mark]

b Efficiency = $\frac{3.9}{10.3} \times 100$ [1 mark]; 38%. [1 mark]

National and global energy resources

1. a Most people are asleep [1 mark] and most appliances are switched off. [1 mark]

b Any two from the following for [1 mark] each: People are waking up; Appliances (e.g. kettles, toasters, showers, lights) are being switched on; Businesses are starting up; People are getting into work.

Section 2: Electricity

Circuit diagrams

1. a Circuit Z. [1 mark]

b Circuit W. [1 mark]

2. Six. [1 mark]

Electrical charge and current

1. a Battery. [1 mark]

b 0.075×10 [1 mark]; 0.75 C [1 mark]

2. a $55 \times 10^{-6} \times 10 \times 60$ [1 mark]; 0.033 C [1 mark]

b $0.048 \div (10 \times 60)$ [1 mark] 8.0×10^{-5} or 0.00008 A [1 mark] which is 80 μA [1 mark]

c 55 + 80 [1 mark]; 135 μA [1 mark]

Electrical resistance

1. $I = \frac{V}{R} = \frac{3.3}{220 \times 10^3}$ [1 mark]; 1.5×10^{-5} A [1 mark]; 15 μA [1 mark]

2. a [1 mark] each for the cell/battery, ammeter and voltmeter connected as shown.

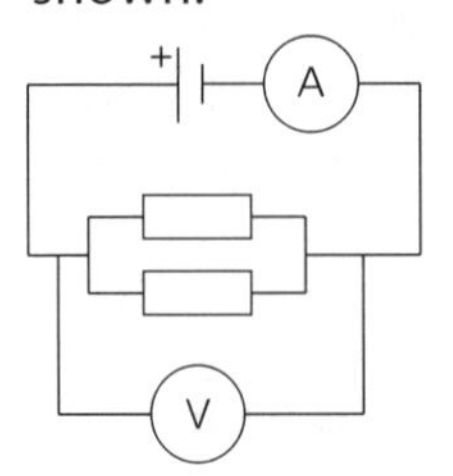

b Divide the voltmeter reading by the ammeter reading (to get the resistance). [1 mark]

3. $I = \frac{V}{R} = \frac{3.00}{12.1}$ [1 mark]; 0.248 A [1 mark]; $Q = It$ $= 0.248 \times 60$ [1 mark]; 14.88 [1 mark]; 14.9 (three significant figures). [1 mark]

4. a Independent variable: length [1 mark]; Dependent variable: resistance [1 mark]; Control variable: diameter/thickness. [1 mark]

b The marks are in three bands according to the level of response.

Level 3 (5/6 marks): A detailed and coherent plan that would produce valid results in sufficient numbers to establish a relationship between resistance and length of wire. All the issues referred to in the question have been addressed and the steps involved are presented in a logical order.
Level 2 (3/4 marks): A detailed and coherent plan with the steps involved presented in a logical order that another person could follow to produce valid results. Not all the issues referred to in the question have been addressed.
Level 1 (1/2 marks): Some simple statements referring to relevant apparatus or steps but may not be in a logical order. The plan would not allow another person to produce valid results.
0: No relevant content.

Indicative content

- Measure the length of wire (connected in the circuit) using a metre rule.
- Use a low value of potential difference to avoid large values of current, which will mean the wire will get hot.
- Temperature affects resistance so you need to control the temperature as far as possible
- Record the ammeter and voltmeter readings.
- Divide the voltmeter reading by the ammeter reading to get the resistance of the wire.
- Open the switch between measurements to prevent the wire from overheating.
- Repeat the procedure for 10 different lengths of wire connected between X and Y.
- Use the full range of lengths of the wire.
- Repeat the measurements for the same lengths of wire.
- Keep the wire fairly tight when measuring it, to improve the accuracy of measurement.
- Check the data to identify any anomalous results, repeat the experiment if anomalies are identified.
- Calculate the mean of the two resistance values for each length to minimise the effect of random errors.

[6 marks]

c Variable on *x*-axis: length [1 mark]; variable on *y*-axis: resistance. [1 mark]

d The best fit line is straight [1 mark] and passes through the origin. [1 mark]

Resistors and I–V characteristics

1. a [2 marks] for each correct resistance value in column four of the table:

Light meter reading (lux)	Current (mA)	Potential difference (V)	LDR resistance (Ω)
1800	75.0	6.0	**80**
150	2.4	6.0	**2500**

b The resistance changes depending on the amount of light that is incident on the LDR for [1 mark]. But stating that the resistance of the LDR gets lower as the brightness increases **or** gets bigger as the brightness decreases for [2 marks].

2. a Between 38 and 42 kΩ [1 mark]; Between 15 and 19 kΩ. [1 mark]

b The resistance changes depending on the temperature of the thermistor for [1 mark] But stating that the resistance gets lower as the temperature increases **or** gets bigger as the temperature decreases for [2 marks].

3. a The current (through an ohmic conductor) is directly proportional to the potential difference across it [1 mark] provided that its temperature is constant. [1 mark]

b $\frac{9}{6} \times 0.80$ [1 mark]; 1.2 A [1 mark]

4. a [1 mark] each for the following four components in the correct positions as shown: switch; ammeter; voltmeter; variable resistor.

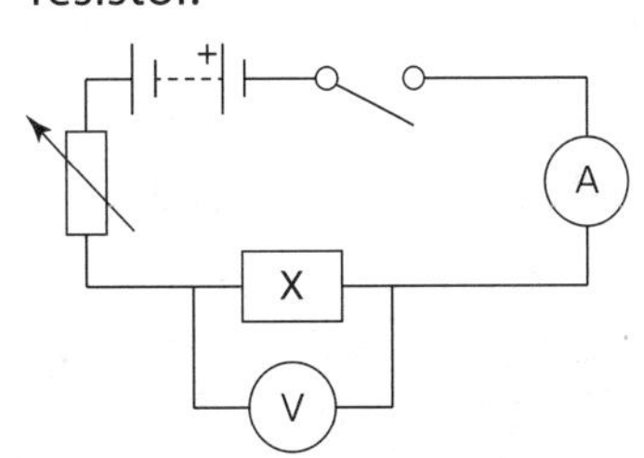

b The marks are in three bands according to the level of response.

Level 3 (5/6 marks): A logically structured explanation of how resistance changes can be deduced from an *I–V* graph, with several comparisons for both changes in size and polarity of the p.d. Nearly all points made are relevant and correct.
Level 2 (3/4 marks): An attempt is made to show how resistance changes can be deduced from an *I–V* graph; some valid comparisons are made. There may be some incorrect or irrelevant points.
Level 1 (1/2 marks): A vague response with one relevant comparison.
0: No relevant content.

Indicative content

- The resistance is found by dividing a potential difference value by the corresponding current value.

• Resistance changes are shown by changes in the gradient. • An increasing gradient indicates that the resistance is decreasing. • As the potential difference increases positively from zero the lamp's resistance increases. • As the potential difference increases positively from zero the diode's resistance decreases. • As the potential difference increases negatively from zero the lamp's resistance increases. • As the potential difference increases negatively from zero the diode's resistance is very high/the diode does not conduct electricity. • The lamp's resistance does not depend on the direction/polarity of the potential difference. The diode's resistance does depend on the direction/polarity of the potential difference.

[6 marks]

Series and parallel circuits

1. Switch open: $V = IR = 0.75 \times 4.0$ [1 mark]; 3.0 V [1 mark]; Switch closed: $V = IR = 1.5 \times 4.0$ [1 mark]; 6.0 V [1 mark]
2. **a** $\frac{1.5}{2000}$ [1 mark]; 0.75 mA [1 mark]
   b 0.75 + 0.75 + 0.75 [1 mark]; 2.25 mA [1 mark]
3. **a** Current through circuit $= I = \frac{V}{R} = \frac{6.0}{(18+12)\times1000}$ [1 mark]; 2.0×10^{-4} (A) [1 mark]
   p.d. across thermistor, $V = IR = 2.0 \times 10^{-4} \times 18 \times 1\,000$ [1 mark]; 3.6 V [1 mark]
   b Current through circuit $= I = \frac{V}{R} = \frac{6.0}{(18+42)\times1000}$ [1 mark]; 1.0×10^{-4} (A) [1 mark]
   p.d. across thermistor, $V = IR = 1.0 \times 10^{-4} \times 18 \times 1\,000$ [1 mark]; 1.8 V [1 mark]

Mains electricity

1. 230 V [1 mark]; 50 Hz [1 mark]
2. The marks are in two bands according to the level of response.

Level 2 (3/4 marks): A coherent description of the sequence of events that lead to an electric shock and a clear explanation of how this is prevented. Correct terminology is used.
Level 1 (1/2 marks): Some relevant points included but may not be in the correct sequence.
0: No relevant content.
Indicative content • The fault creates a potential difference between the metal casing and the ground/earth. • If a person touches the casing, their body completes the circuit. • Without an earth wire, an electric current could flow through the person's body to the ground giving the person an electric shock. • The earth wire is connected to the casing of the toaster. • When the fault occurs, electric current flows along the earth wire to the ground instead of through the person… • Because it has a very low resistance; • …protecting the person from electric shock/preventing the current passing through the person.

[4 marks]

Energy changes in circuits

1. The power of the appliance [1 mark]; the length of time that it is switched on. [1 mark]
2. $1200 \times 10 \times 60$ [1 mark] = 720 000 J [1 mark];
3. **a** $1.5 \times 6 \times 3600 \times 365$ [1 mark]; 1.2×10^7 J [1 mark]; expressed in standard form to two significant figures as shown gets an additional mark.
   b $\frac{1.2\times10^7}{3.6\times10^6} \times 15$ [1 mark]; 49p [1 mark]
4. For PC, $100 \times 30 \times 60 = 1.8 \times 10^5$ J [1 mark]; time $= \frac{\text{energy}}{\text{power}}$ [1 mark] $= \frac{1.8\times10^5}{12} = 1.5 \times 10^4$ s [1 mark] = 250 min [1 mark]
5. **a** $2500 \times 50 \times 60$ [1 mark] $= 7.5 \times 10^6$ J or 7 500 000 [1 mark]; answer given in standard form as shown. [1 mark]
   b $\frac{7.5\times10^6}{230}$ [1 mark]; 3.26×10^4 C or 32 608 [1 mark]; 3.3×10^4 C (to two significant figures in standard form). [1 mark]

Electrical power

1. Current $I = \frac{P}{V} = \frac{1100}{230}$ [1 mark] = 4.78(26) A [1 mark]; 4.8 (two significant figures) [1 mark]

2. Current $I = \frac{P}{V} = \frac{2300}{230}$ [1 mark] = 10 A [1 mark];
   $R = \frac{V}{I} = \frac{230}{10}$ [1 mark]; 23 Ω [1 mark].
3. $(3.0)^2 \times 1$ [1 mark]; 9.0 W [1 mark]

The National Grid

1. **a** Power dissipated by a 100 A current: $P = I^2R = 100^2 \times 0.1$ [1 mark] = 1000 W [1 mark]; Power dissipated by a 1000 A current: $P = I^2R = 1000^2 \times 0.1$ [1 mark] = 100 000 W [1 mark]; More power dissipated by the 1000 A current. [1 mark]

   But: a 10 times larger current dissipates 100 times more power for [2 marks].

   b Small current/high p.d. [1 mark]
2. **a** A step-up transformer increases potential difference but a step-down transformer decreases potential difference. [1 mark]

   b Transformer A: step-up [1 mark]; Transformer B: step-down. [1 mark]

Static electricity

1. **a** Electrons are transferred [1 mark] from the rod to the cloth. [1 mark]

   b The cloth becomes negatively charged. [1 mark]

   c The student charges the second acetate rod and the polythene rod with the cloths [1 mark]; He moves the acetate rod towards the suspended rod and observes the rods repel each other [1 mark]; He then moves the polythene rod towards the suspended rod and observes the rods attract each other. [1 mark]

Electric fields

1. **a** Negative charge. [1 mark]

   b The speck of dust would be repelled/move away from the sphere. [1 mark]
2. **a** Repulsive. [1 mark]

   b The force gets bigger. [1 mark]

Section 3: Particle model of matter

Density

1. **a** The marks are in two bands according to the level of response.

Level 2 (3–4 marks): Evidence set out clearly to reach a conclusion. Correct density calculated for the sample and correct conclusion drawn.
Level 1 (1–2 marks): Use of an equation and an attempt to compare to one of the given densities. Response may not be coherently structured.
Indicative content • Volume of block = $8.0 \times 5.0 \times 2.0 = 80$ cm^3. • Density = $\frac{M}{V} = \frac{624}{80} = 7.8$ g/cm^3. • Convert density to 7800 kg/m^3. • Comparing with density table suggests the block is made of steel.

[4 marks]

   b Could be other metals (not in the table) that have densities around 7800 kg/m^3 [1 mark]; [1 mark] for **one** of the following: Don't know how accurate the density value actually is; Don't know the uncertainty in the density value; Could be tin if the uncertainty in the density is large enough.

   c Any **one** of the following for [1 mark]: ±1mm, ±0.1 cm, ±0.5 mm.
2. **a** The marks are in two bands according to the level of response.

Level 2 (3/4 marks): The plan is detailed and coherent. The steps are in an order that makes sense and could be followed by someone else to obtain valid results.
Level 1 (1/2 marks): Basic description of measurements needed with no indication of how to use them. Would be difficult for another person to follow.
0: No relevant content.
Indicative content • Measure the mass of the pebble using the electronic balance. • Put one empty beaker under the spout of the displacement can. • Use the second beaker to pour water into the displacement can until water comes out of the spout. • When the water has stopped dripping place the (dry) measuring cylinder under the spout.

- Lower the pebble (tied on a string) carefully into the displacement can so that it is completely submerged.
- Collect all the water that comes out of the spout in the measuring cylinder.
- Record the volume of water in the measuring cylinder.
- The volume of water (displaced) is equal to the volume of the pebble.
- Calculate the density of the pebble using density = $\frac{\text{mass}}{\text{volume}}$.

[4 marks]

b $\frac{84.2}{38.5}$ [1 mark]; 2.19 g/cm^3 [1 mark]

c The wider measuring cylinder has graduations corresponding to a larger volume than the narrower measuring cylinder. [1 mark]. This gives a bigger uncertainty in the volume of the pebble. [1 mark]. So this gives a bigger uncertainty in the density measurement. [1 mark]

Changes of state

1. Evaporation [1 mark]; Sublimation [1 mark]

2. In a solid state the particles can only vibrate about a fixed position but in the liquid state, the particles are able to move around [1 mark]; The particles in a solid are in a regular pattern/arrangement but in the liquid state they have no regular pattern [1 mark]; The particles are close together in both the solid and liquid states. [1 mark]

3. a 10–20 minutes [1 mark]

b (Approximately) 70 °C [1 mark]

Internal energy and specific latent heat

1. a $\Delta E = m\,c\,\Delta\theta = 1000 \times 500 \times (1\,200 - 20)$ [1 mark]; 5.9×10^8 J [1 mark]; 590 MJ [1 mark]

b $E = m\,L = 1000 \times 10\,000$ [1 mark] $= 1.0 \times 10^7$ J [1 mark]; 10 MJ [1 mark]

c Total energy = 590 + 10 [1 mark]; 600 MJ [1 mark]

2. a The kinetic energy of all the atoms/molecules [1 mark]; The potential energy of all the atoms/molecules. [1 mark]

b The kinetic energy of the atoms/molecules. [1 mark]

3. a 50 °C [1 mark]; 150 °C [1 mark]

b Specific heat capacity in the liquid state is greater [1 mark]; The temperature in the liquid state increases more slowly (than in the solid state). [1 mark];

4. Mass of water changed to steam = 581 – 526 = 55 g = 0.055 kg [1 mark]

Energy supplied = 327 – 195 = 132 kJ = 132 000 J [1 mark];

$L = \frac{E}{m} = \frac{132000}{0.055}$ [1 mark]; 2.4×10^6 J/kg [1 mark]; give an additional mark if the answer is given in standard form as shown.

Particle motion in gases

1. a When molecules collide with the container walls [1 mark]; They exert forces on the walls. [1 mark]

b Faster molecules exert bigger forces on the container walls [1 mark]; Faster molecules hit the walls more frequently. [1 mark]

2. a 120–125 kPa [1 mark]

b 190–210 kPa [1 mark]; line on the graph shown extrapolated [1 mark]; Assume that the gas follows the same pattern of expansion at the higher temperatures. [1 mark]

Increasing the pressure of gas

1. a The molecules are further apart [1 mark]; The number of collisions with the container walls are less frequent [1 mark]; The gas pressure decreases. [1 mark]

b pV = constant $= 0.18 \times 2.5 \times 10^5$ [1 mark]; 4.5×10^4 [1 mark];

$p_{new} = \frac{4.5\times10^4}{0.25}$ [1 mark]; 1.8×10^5 Pa [1 mark]; Give an additional mark for the answer given in standard form as shown.

2. The marks are in two bands according to the level of response.

Level 2 (3/4 marks): A clear coherent explanation that considers temperature, pressure and internal energy of the air.
Level 1 (1/2) Some relevant points made; lacks detail; may have considered only two from: temperature, pressure and internal energy.
0: No relevant content.

Indicative content
• The work done increases the speed of the molecules resulting in an increase in temperature. • The increased speed of the molecules results in an increase in their kinetic energy and therefore an increase in internal energy. • The increased speed of the molecules results in an increase in the force they exert on the walls of the barrel and therefore an increase in pressure. • The work done reduces the volume of the air and so increases the frequency of the collisions of the molecules with the walls of the barrel increasing the pressure.

[4 marks]

Section 4: Atomic structure

Protons, neutrons and electrons

1. a See diagram: neutron correctly labelled [1 mark]; electron correctly labelled. [1 mark]

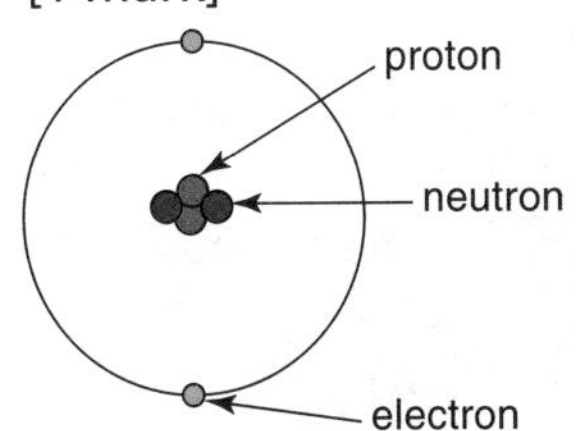

b Protons and electrons are charged [1 mark for both]; protons are positive [1 mark]; electrons are negative. [1 mark]

2. a Mass = $12 \times 1.7 \times 10^{-27}$ [1 mark] = 2.0×10^{-26} kg [1 mark]; give an additional mark if the answer is given to two significant figures in standard form as shown.

b $\frac{1.7\times10^{-27}}{9.1\times10^{-31}}$ [1 mark]; 1900 [1 mark]; If the answer 1 868 is given then maximum [1 mark].

The size of atoms

1. a Times bigger = 10^4 [1 mark]; orders of magnitude = 4 [1 mark]

b Number of times bigger = $\frac{6\times10^{-11}}{3\times10^{-15}}$ [1 mark]; 20 000 [1 mark]

Elements and isotopes

1. The marks are in two bands according to the level of response:

Level 2 (3/4 marks): A clear, detailed and coherent comparison of the isotopes using correct scientific terminology and including both similarities and differences.
Level 1 (1/2 marks): Some relevant details of the isotopes listed but comparisons may not made.
0: No relevant content.
Indicative content • The three isotopes have the same number of protons in their nuclei. • The three isotopes have same number of orbiting electrons. • The isotopes have different numbers of neutrons in their nuclei. • $^{14}_{6}C$ has 8 neutrons in its nucleus; $^{13}_{6}C$ has 7 and $^{12}_{6}C$ has 6. • (Of the three isotopes) an atom of $^{14}_{6}C$ has the largest mass and $^{12}_{6}C$ has the smallest mass.

[4 marks]

2. a $^{197}_{79}Au$ [1 mark] for 79 in correct place; [1 mark] for 197 in correct place.

b 79 [1 mark]

c $^{205}_{79}Au$ has 205 nucleons.

Mass of nucleus of $^{205}_{79}Au$ = $(205 \times 1.7 \times 10^{-27})$ [1 mark] = 3.485×10^{-25} kg [1 mark]; $^{169}_{79}Au$ has 169 nucleons.

Mass of nucleus of $^{169}_{79}Au$ = $(169 \times 1.7 \times 10^{-27})$ [1 mark] = 2.873×10^{-25} kg [1 mark]; Number of times more massive = $\frac{3.485\times10^{-25}}{2.873\times10^{-25}}$ = 1.2(13) [1 mark]; 1.2 (two significant figures). [1 mark]

Electrons and ions

1. a Nucleus: $11 \times 1.6 \times 10^{-19}$ [1 mark]; (+) 1.8×10^{-18} C [1 mark]; Electrons: -1.8×10^{-18} C [1 mark]; (no second mark for repeating the calculation); two significant figures for both answers [1 mark]; atom: 0 C. [1 mark]

b + [1 mark]; 1.6×10^{-19} C [1 mark]

2. (+)1.6×10^{-19} C [1 mark]. There are three protons and two electrons [1 mark]; so the overall charge is equal to the charge on one proton. [1 mark]

Discovering the structure of the atom

1. **a** Either a significant number of alphas were scattered through large angles or some alphas were scattered backwards. [1 mark]

   b The marks are in two bands according to the level of response.

Level 2 (3/4 marks): A detailed coherent comparison is provided in which differences/similarities are clearly identified.
Level 1 (1/2 marks): Some relevant statements included but comparisons are not made.
0: No relevant content.
Indicative content • Mass is distributed throughout the atom in the plum pudding model but is concentrated in the centre of the atom/central nucleus in the nuclear model. • Both models have positive and negative charge present in the atom. • Positive charge is distributed throughout the atom in the plum pudding model but is contained within the central nucleus in the nuclear model. • Negative charge is embedded in the positive charge in the plum pudding model but is separated from the positive charge in the nuclear model.

[4 marks]

2. The marks are in three bands according to the level of response.

Level 3 (5/6 marks): A clear, detailed and coherent description is provided.
Level 2 (3/4 marks): A basic description is provided. There may be some incorrect or irrelevant points.
Level 1 (1/2 marks): Simple statements given. The descriptions are vague and lack sufficient detail.
0: No relevant content.
Indicative content • Most of the mass is located in a tiny central nucleus. • All the positive charge is located in the nucleus. • The negative charge is carried by the orbiting electrons. • The electrons orbit in specific energy levels/shells. • The particles in the nucleus are neutrons and protons. • Most of the atom is empty space. • Protons are positively charged particles. • Neutrons have no charge. • Protons and neutrons have the same/similar mass. • Protons and electrons have the same magnitude of charge but are oppositely charged. • A proton is about 2000 times more massive than an electron. • Typical radius of an atom is 1×10^{-10} m.

[6 marks]

Radioactive decay

1. **a** Electron [1 mark]

   b The uranium nucleus has one more proton [1 mark], and one fewer neutron. [1 mark]

2. **a** Atomic number: 28 [1 mark]; mass number: 60. [1 mark]

   b Electromagnetic. [1 mark]

Comparing alpha, beta and gamma radiation

1. **a** $120 \div 300$ [1 mark]; 0.4 [1 mark]

   b 2230 [1 mark]

   c All values in table correct (see below) [1 mark]; all values to one decimal place. [1 mark]

Absorber (mm)	Number of counts	Counting time (s)	Count-rate in counts per second
1	1601	100	16.0
2	822	100	8.2
3	401	100	4.0
4	399	100	4.0

d Range: 3 mm [1 mark]; the 2 mm reading is above background [1 mark]; the 3 mm reading is either just due to background radiation or shows that all the beta particles have been stopped. [1 mark]

2. a [1 mark] for each range (see table below).

Radiation	Range
Alpha	A few cm
Beta	A few metres
Gamma	Not affected by air

b Alpha. [1 mark]

Radioactive decay equations

1. ${}^{241}_{95}$Americium = ${}^{237}_{93}$Neptunium + ${}^{4}_{2}$He; [1 mark] each for the correct mass number and atomic number for neptunium; correct symbol for the alpha particle. [1 mark]
2. ${}^{28}_{13}$Al = ${}^{28}_{14}$Si + ${}^{0}_{-1}$e [1 mark] each for the correct mass number and atomic number for silicon; correct symbol for the beta particle. [1 mark]
3. a Beta [1 mark]

   b ${}^{27}_{12}$Mg = ${}^{27}_{13}$Al + ${}^{0}_{-1}$e ; [1 mark] for ${}^{27}_{13}$Al and [1 mark] for ${}^{0}_{-1}$e

   c The nucleus emits either a gamma ray or electromagnetic radiation. [1 mark]

Half-lives

1. a Fraction remaining = $\frac{480}{1920} = \frac{1}{4}$ [1 mark]

   b Two half-lives have passed [1 mark]; half-life = 15 hours. [1 mark]
2. a 48 days = 3 half-lives [1 mark]; $\frac{1}{8}$ remaining [1 mark]; 0.15 mg. [1 mark]

   b $\frac{1}{8} \times 3600$ [1 mark]; 450 Bq. [1 mark]
3. a Beta [1 mark]

   b 32 days = 4 half-lives [1 mark]; $\frac{1}{16}$ remaining [1 mark]; 750 Bq [1 mark]
4. a The outcome cannot be predicted. [1 mark]

   b 70% [1 mark]

   c 48% [1 mark]

   d There is no (predictable) ratio/percentage of heads/tails when throwing a very small number of coins [1 mark]. Throwing a larger number of coins, the predicted ratio/percentage of half showing heads and half showing tails starts to appear. [1 mark]

   e There is a very large number of atoms (even in a sample with a small mass). [1 mark]
5. a Suitable best fit curve passing through all points (within 1 small grid square). [1 mark]

   b Two half-life values in the range 45 to 55 s [2 marks]; The determination of each value shown on the graph [2 marks]; Mean value in the range 45 to 55 s. [1 mark]

Radioactive contamination

1. a The unwanted presence of radioactive atoms. [1 mark]

   b Absorbed by the skin [1 mark]; Breathed in [1 mark]; via an open wound. [1 mark]

   c Prevents transfer (of radioactive atoms) to the workers' skin [1 mark]; lungs [1 mark]; and their own clothing. [1 mark]
2. Gamma radiation is very penetrating and would pass through the protective clothing. [1 mark]

Background radiation

1. a Radon [1 mark]

   b Medical [1 mark]

   c Radon is a gas and therefore can be breathed in [1 mark]. Alpha radiation is the most ionising nuclear radiation. [1 mark]
2. a The types of rock that properties are built on varies across the UK [1 mark]. Rocks used as raw materials in construction may contain different amounts of radioactive isotopes. [1 mark]

   b Cosmic rays are more intense at higher altitude [1 mark]; at altitude there is less atmosphere above you to absorb the cosmic rays. [1 mark]

Uses and hazards of nuclear radiation

1. a Alpha radiation would not be able to travel through the soil (at the location of the leak) to be detected/alpha radiation not very penetrating. [1 mark]

   b The sodium-24 is active long enough to find the leak [1 mark] but the activity decreases quickly enough so as not to harm living things in the environment. [1 mark]

2. The marks are in two bands according to the level of response.

Level 2 (3/4 marks): A detailed and clear explanation involving several factors, logically presented.
Level 2 (1/2 marks): Some relevant points included; may not have used scientific terminology.
0: No relevant content.
Indicative content • The 6-hour (short) half-life means that the isotope remains active long enough to carry out the investigation. • But the activity decreases quickly enough so harm to the patient is minimised. • Gamma radiation is very penetrating so it can pass through the patient's body to the gamma camera. • Gamma rays have low ionising power. • Low energy gamma rays produce less ionisation causing less harm to the patient.

[4 marks]

Nuclear fission

1. a At least one of the neutrons created by fission of a uranium-235 nucleus [1 mark] goes on to induce another uranium-235 nucleus to undergo fission (and so on). [1 mark].
   b Kinetic energy [1 mark] of the fission fragments and the (free) neutrons. [1 mark]
   c The products of the fission reactions collide with other atoms in the reactor core [1 mark] and transfer kinetic energy to these atoms (raising the temperature). [1 mark]
   d Control rods are made of a material that absorbs neutrons [1 mark]. The rods are lowered into the reactor to slow down (or halt) the chain reaction (due to the decrease in the number of neutrons (produced from fission) available to cause further fission. [1 mark]
2. The marks are in two bands according to the level of response.

Level 2 (3/4 marks): A coherent description involving half-life, relative levels of activity as time passes and the hazard created.
Level 2 (1/2 marks): Some relevant statements included but but the link with decreasing activity levels is not shown.
0: No relevant content.
Indicative content • Barium-140 has the shortest half-life and so at first has the highest activity. • In a short time (few months) the number of barium-140 nuclei has decreased to almost zero so the barium-140 is no longer a hazard. • Over the next few hundred years both strontium-90 and caesium-135 are active but the strontium-90 has the shortest half-life and so contributes most to the activity (of the waste) and presents the biggest hazard. • After about 1000 years, the strontium-90 will have disappeared (almost no radioactive nuclei remaining) so its activity is zero and the waste's activity will be due to the caesium. • The activity will now be low and the hazard level is small.

[4 marks]

Nuclear fusion

1. a Protons are positively charged [1 mark] and so repel each other. [1 mark]
   b A proton changes into a neutron. [1 mark]
   c Atomic number = 2 [1 mark]; mass number = 3 [1 mark]
   d $^{1}_{1}$hydrogen + $^{2}_{1}$hydrogen = $^{3}_{2}$helium; [1 mark] for each isotope, maximum [3 marks]
   e $^{3}_{2}He + ^{3}_{2}He = ^{4}_{2}He + \left(2 \times ^{1}_{1}H\right)$; [1 mark] for each part of the equation, maximum [4 marks]

Section 5: Forces

Scalars and vectors

1. a 20 km [1 mark]
   b Distance: an answer in the range 9–11 km [1 mark]; Direction: southeast. [1 mark]
2. Distance: in the range 19–21 m [1 mark]; Direction: northeast. [1 mark]

Speed and velocity

1. $\frac{60}{6.4}$ [1 mark]; 9.375 m/s [1 mark] 9.4 to 1 sig fig [1 mark]; $\frac{9.375 \times 3600}{1000}$ [1 mark]; 33.75 [1 mark] 34 km/h (two significant figures). [1 mark]
2. 300 s. [1 mark]
3. **a** 1 mark for each correct speed in column An additional mark if data all given to two significant figures as shown.

Stage	Time (minutes)	Average speed (m/s)
1500 m swim	25	1.0
40 km cycle	70	9.5
10 km run	42	4.0

b Average speed = $\frac{(1.5+40+10)\times 1000}{(25+70+42)\times 60}$ [1 mark]; 6.26 (5) [1 mark] 6.3 m/s (two significant figures). [1 mark]

c Each stage should be shown as a straight line indicating the correct distance travelled and time interval. [1 mark] for a straight line in each of the three sections: Swimming from zero to 1.5 km at 25 minutes; cycling to 41.5 km at 95 minutes; running to 51.5 km at 137 minutes.

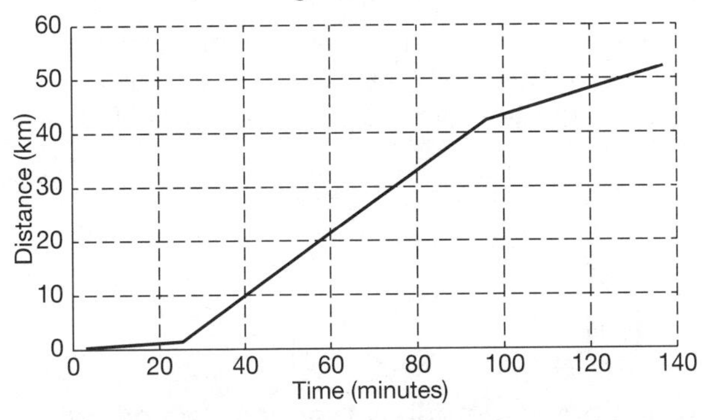

4. Drawing a line segment to the line at 3s [1 mark]; gradient triangle shown on graph [1 mark]; 20 m/s (allow range 18.0 to 22.0). [1 mark]

Acceleration

1. Converting speeds to m/s: 22 km/h = $22 \times \frac{1000}{3600} = 6.11$ m/s [1 mark]; 28 km/h $= 28 \times \frac{1000}{3600} = 7.78$ m/s [1 mark]; acceleration $a = \frac{\Delta v}{t} = \frac{7.78-6.11}{8.0}$ [1 mark] = 0.208 [1 mark] 0.21 m/s^2 (two significant figures) [1 mark]
2. Gradient triangle drawn on graph [1 mark]; 5 m/s^2 [1 mark]
3. Attempt to draw the tangent to the curve at $t = 10$ s [1 mark]; Gradient triangle shown on graph [1 mark]; Accept a value from 1.8 to 2.4 m/s [1 mark]
4. Evidence of an area calculation, see below [1 mark]; Displacement = (5 × 20) + 20 + 30 [1 mark] = 150 m [1 mark]

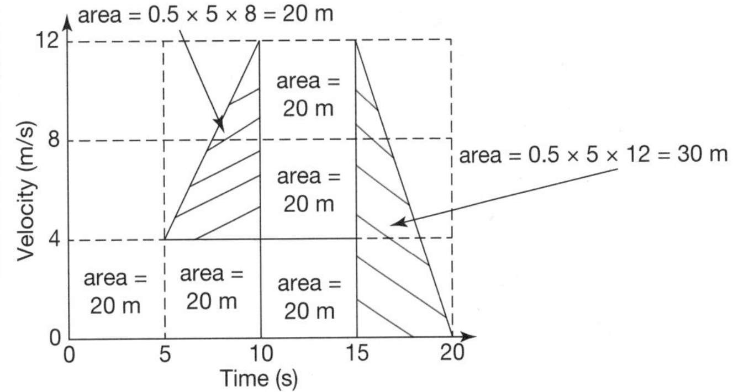

Equation for uniform acceleration

1. $v = \sqrt{2as}$ or $v^2 = 2as - 0$ [1 mark]; $\sqrt{2 \times 9.8 \times 150}$ [1 mark]; 54.2(2) [1 mark]; 54 m/s (two significant figures). [1 mark]
2. $v^2 = u^2 + 2as$ [1 mark]; $v^2 = 4.0^2 + (2 \times 0.40 \times 25)$ [1 mark]; $v = 6.0$ m/s [1 mark]
3. $a = \frac{v^2}{2s}$ [1 mark]; $\frac{2.5^2}{2 \times 2.0}$ [1 mark]; 1.56(2) [1 mark] 1.6 m/s^2 (two significant figures). [1 mark]

Forces

1. **a** [1 mark] for each labelled arrow (see diagram below).

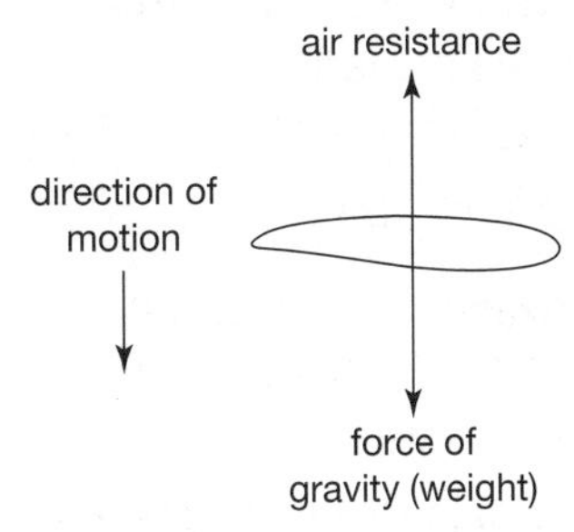

b Air resistance is a contact force [1 mark]; weight is a non-contact force. [1 mark]

2. [1 mark] for each correctly labelled arrow (see diagram below).

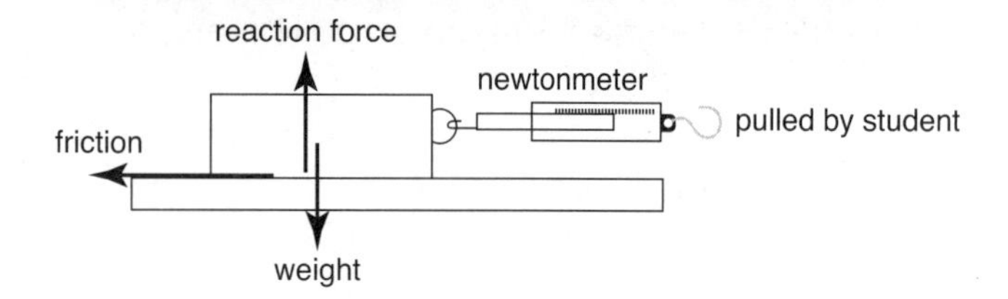

b Before the air is removed, the force exerted by the gas molecules on the inside and outside surfaces are balanced [1 mark]; The vacuum pump removes the air from inside the can [1 mark]; The force on the outside of the can is no longer balanced and crushes the can. [1 mark]

Moment of a force

1. 10×0.76 [1 mark]; 7.6 N m [1 mark]
2. a $1.0 \times (50 - 12)$ [1 mark]; 38 N cm [1 mark]
   b Rearrange to give distance $d = \frac{\text{moment}}{\text{force}}$ [1 mark]; $d = \frac{38}{2.0}$ [1 mark]; 19 cm. [1 mark]
   c 69 cm. [1 mark]

Levers and gears

1. a Force $F = \frac{\text{moment}}{\text{distance}}$ [1 mark]; $\frac{1000 \times 0.40}{1.60}$ [1 mark]; 250 N. [1 mark]
   b Max weight $= \frac{\text{moment}}{\text{distance}}$ [1 mark] $= \frac{360 \times 1.60}{0.40}$ [1 mark]; 1.44 kN [1 mark] 1.4 kN (two significant figures). [1 mark]
2. Max. weight $= \frac{\text{moment}}{\text{distance}}$ [1 mark] $= \frac{360 \times 1.3}{0.40}$ [1 mark]; 1.17 kN [1 mark]; 1.2 kN (two significant figures). [1 mark]
3. Smaller gear/driven gear. [1 mark].

Pressure in a fluid

1. $p = \frac{F}{A}$ so units are $\frac{N}{m^2}$ [1 mark]; N/m^2 [1 mark]
2. a Force is at right angles/90° to a surface; [1 mark]
   b Rearrange to give $F = pA$ [1 mark]; $1.01 \times 10^5 \times 1.14$ [1 mark]; 1.1514×10^5 N [1 mark]; 1.15×10^5 N (three significant figures). [1 mark]
3. $p = h\rho g$ [1 mark]; $= 50 \times 1.0 \times 10^3 \times 9.8$ [1 mark]; 4.9×10^5 Pa [1 mark]

Atmospheric pressure

1. a Pressure in the range 35 000–38 000 Pa. [1 mark]
   b Altitude: 4 000 m. [1 mark]
2. a Molecules (of gas) in the air collide with a surface [1 mark]; These molecules exert a force (when they collide) with a surface. [1 mark]

Gravity and weight

1. The mass has the same value on the Earth and the Moon [1 mark]; The weight is greater on the Earth than on the Moon. [1 mark]
2. Mass of fuel $= \rho V$ [1 mark]; 800×0.50 [1 mark]; 400 kg [1 mark]; Change in weight $= 400 \times 9.8$ [1 mark] $= 3920$ N (accept 3900). [1 mark]

Resultant forces and Newton's first law

1. $11.8 - 5.1 = 6.7$ N [1 mark]; towards the first child. [1 mark]
2. a Inertia. [1 mark]
   b Air resistance arrow vertically upwards [1 mark]; weight (or force or gravity) arrow vertically downwards [1 mark]; arrows about the same size. [1 mark]
3. Two vectors drawn the same length (see diagram below) [2 marks]; Resultant drawn [1 mark]; Resultant in the range 6 to 8 N. [1 mark]

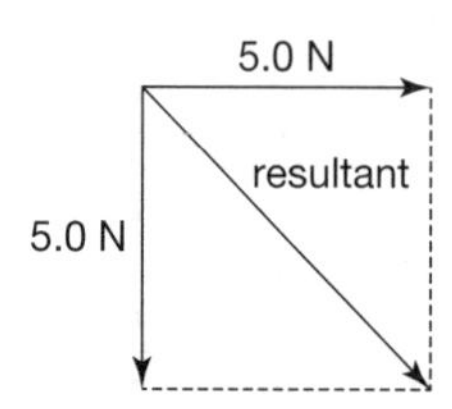

4. a Components drawn correctly (see below). [1 mark]

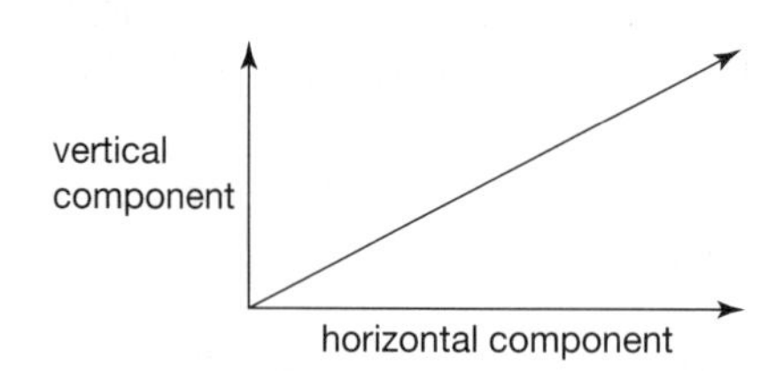

   b Horizontal component in the range 8.5 to 9.5 N [1 mark]; vertical component in the range 4.5 to 5.5 N. [1 mark]

Forces and acceleration

1. a The air track enables the glider to move on a cushion of air reducing friction [1 mark]; The glider has a streamlined shape to reduce air resistance. [1 mark]
   b The glider should be moved to the left of the air track and given a gentle push [1 mark]; The glider is travelling at a steady speed if the readings on the timers connected to both light gates have the same value. [1 mark]
   c The height of air track above the bench can be adjusted to make the right hand end slightly lower. [1 mark]
   d Initial velocity $u = 0.100 \div 1.78$ [1 mark] = 0.0562 m/s [1 mark]; Final velocity $v = 0.100 \div 0.52$ [1 mark] = 0.192 m/s [1 mark]; $a = \frac{0.192^2 - 0.0562^2}{2 \times 0.5}$ [1 mark]; 0.034 m/s^2 [1 mark]
   e The graph would be a straight line [1 mark] through the origin. [1 mark]
2. Rearrange to give acceleration = $F \div m$ [1 mark]; $2700 \div 2400$ [1 mark]; 1.125 [1 mark]; 1.1 m/s^2 (two significant figures). [1 mark]
3. a Rearrange to give $a = \frac{v^2 - u^2}{2s}$ [1 mark]; $\frac{8.2^2 - 4.0^2}{2 \times 40}$ [1 mark]; 0.640(5); [1 mark] 0.64 m/s^2 (two significant figures). [1 mark]
   b Resultant force = 63×0.64 [1 mark]; 40 N or 40.32 N [1 mark]; Driving force = 60 N (accept 60.32 N). [1 mark]
4. a 10 km/h. [1 mark]
   b In the range 14 to 15 N. [1 mark]

Terminal velocity

1. a Acceleration is decreasing. [1 mark]
   b Zero. [1 mark]
   c See diagram below. [1 mark] each for A [1 mark] and for B. [1 mark] for C provided the two arrows are the same length. Give an additional mark if the three weight arrows are all the same length.

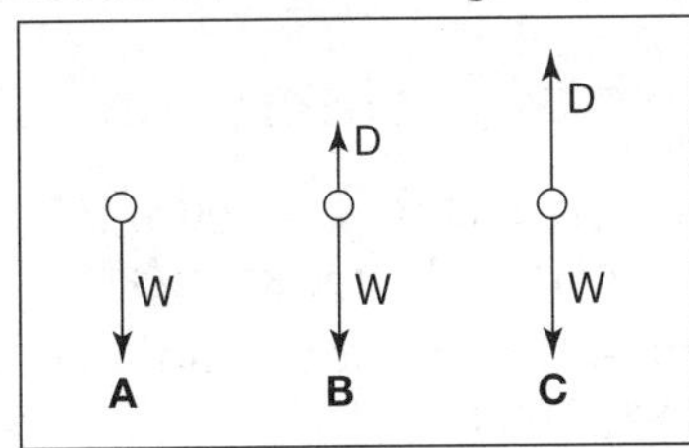

2. a An attempt to determine the gradient [1 mark]; Acceleration in the range 0.8–1.2 m/s^2 [1 mark]
   b Weight of parachutist = $mg = 70 \times 9.8$ [1 mark]; 686 N [1 mark]; Resultant force = ma = 70×1 [1 mark] = 70 N [1 mark]; Resultant force = weight – air resistance [1 mark]; Air resistance = 686 – 70 = 616 N [1 mark]; acceptable range 630–600 depending on the value obtained for acceleration.

Newton's third law

1. a Upwards [1 mark]; gravitational (attraction)/ force of gravity. [1 mark] the Earth (do not accept the ground) [1 mark];
   b Downwards [1 mark]; normal contact force. [1 mark]; the ground [1 mark];
2. a Elastic potential energy store of the balloon. [1 mark]
   b The stretched balloon pushes air out of the neck of the balloon. [1 mark]; The air pushes back on the balloon in the opposite direction with an equal force. [1 mark]

Work done and energy transfer

1. a $0.50 \times 9.8 \times 1.0$ [1 mark]; 4.9 J [1 mark]
   b $\frac{4.9}{2.1}$ [1 mark]; 2.33(3) W [1 mark]; 2.3 W (two significant figures). [1 mark]
2. a $50\,000 \times 800 \times 10^3$ [1 mark]; 4.0×10^{10} J [1 mark]; 40 GJ [1 mark]
   b $\frac{4.0 \times 10^{10}}{3600}$ = [1 mark]; $1.11(1) \times 10^7$ W [1 mark]; 11.1(1) MW [1 mark]; 11 MW (two significant figures). [1 mark]

Stopping distance

1. a 72 km/h = $\frac{72 \times 1000}{3600}$ = 20 m/s [1 mark]; kinetic energy = $\frac{1}{2}mv^2 = 0.5 \times 800 \times 20^2$ [1 mark] = 1.6×10^5 J [1 mark]
   b 20×0.60 [1 mark] = 12 m [1 mark]
   c Braking distance = $\frac{\text{work done}}{\text{braking force}}$ [1 mark] = $\frac{1.6 \times 10^5}{5000}$ [1 mark] = 32 m [1 mark]
   d 12 + 32 = 44 m [1 mark]
2. a 32 km/h = 8.9 m/s [1 mark]; rearrange to give reaction time = $\frac{\text{distance}}{\text{time}}$ [1 mark] = $\frac{6.0}{8.9}$ [1 mark] = 0.67 s [1 mark]

b Two of the following for [1 mark] each: tiredness; distractions; drugs/alcohol.

3. a Values read from graph 20 + 70 [1 mark]; 90 m [1 mark]

b $\frac{1}{2}mv^2 = F\,d$ [1 mark]; rearrange to give $F = \frac{1}{2}mv^2 \div d$ [1 mark] $= \frac{1}{2} \times 810 \times 30^2 \div 70$ [1 mark]; 5200 N [1 mark]

4. Attempt to read values from graph [1 mark]; Percentage increase $= \frac{(33-24)}{24} \times 100$ [1 mark]; accept in range 36 to 38 % [1 mark]

Force and extension

1. a Diagram (see below) shows a 90° set square [1 mark] aligned to the desk and the rule [1 mark]. Alternatively a plumb line could be shown alongside the metre rule [2 marks].

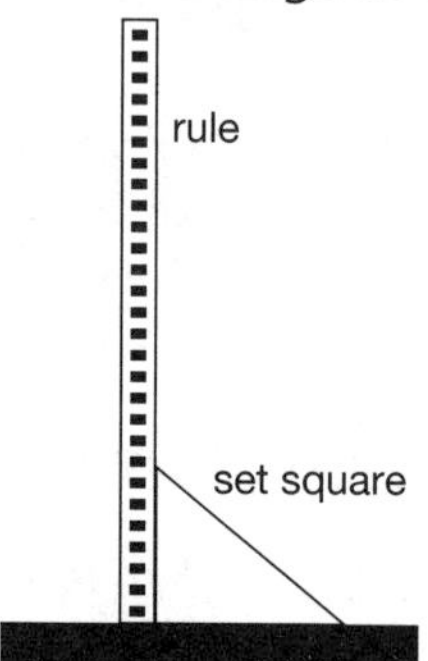

b The marks are in three bands according to the level of response.

Level 3 (5/6 marks): The plan is detailed and coherent with all steps described in the correct logical sequence including how the data should be analysed. The issue of random errors is addressed.
Level 2 (3/4 marks): The plan is detailed and coherent with all major steps described in the correct logical sequence including how the data should be analysed. Some detail may be missing.
Level 1 (1/2 marks): Some relevant steps included. May not be presented in a logical sequence.
0: No relevant content.
Indicative content • The initial length of the spring with no weights attached is measured (using the metre rule). • A 1.0 N weight is attached to the spring and the length of the spring is measured. • The extension of the spring is found by subtracting the initial length of the spring from the length with the weight attached. • The extension values are found for weights up to 8.0 N. • The weights attached to the spring are decreased to enable repeat measurements for extension to be determined. • The average extension value for each weight is calculated. • A graph of extension on the *y*-axis versus weight on the *x*-axis is plotted.

[6 marks]

2. a 0 (or 1) to 14 N (accept 13 to 15 N). [1 mark]

b Gradient of straight line (triangle drawn on graph) [1 mark] = 0.014 [1 mark]; spring constant $= \frac{1}{\text{gradient}}$ [1 mark] 71 [1 mark]; N/m [1 mark]

c 64.3 $E_e = \frac{1}{2}ke^2 = 0.5 \times 71 \times 0.14^2$ [1 mark]; 1.4 J [1 mark]

3. $E_e = \frac{1}{2}ke^2 = 0.5 \times 20 \times 0.1^2$ [1 mark] = 0.10 J [1 mark]. Assume all the elastic potential energy stored in the spring is converted to gravitational potential energy of the ball [1 mark]; rearrange to give height $= \frac{\text{energy}}{mg}$ [1 mark] $= \frac{0.10}{0.003 \times 9.8}$ [1 mark]; 3.401 [1 mark] 3.4 m (two significant figures). [1 mark]

Momentum

1. Velocity 20 km/h $= \frac{20 \times 1000}{3600}$ = 5.56 m/s [1 mark]; velocity 40 km/h = 11.11 m/s [1 mark];
Momentum change = 3 000 × (11.11 – 5.56) [1 mark]; 1.7×10^4 kg m/s [1 mark]

2. a Momentum = 0.057 × 15.0 [1 mark]; 0.855 kg m/s [1 mark]

b Momentum = –0.855 kg m/s [1 mark for the minus sign] [1 mark for the value]

c Momentum change 1.71 kg m/s [1 mark]

Conservation of momentum

1. Momentum before collision = (800 × 10) + (1000 × 6.0) [1 mark] = 14 000 kg m/s [1 mark]
Momentum after the collision = momentum before collision [1 mark]
$800v_{car} + (1000 \times 8.0) = 14\,000$ [1 mark]

Velocity of the car: $v_{car} = \frac{14000-8000}{800} = 7.5$ m/s [1 mark]

2. a $p = mv = 1.7 \times 10^{-27} \times 2.70 \times 10^{7}$ [1 mark]; 4.59×10^{-20} kg m/s [1 mark]
   b $1.83 \times 10^{-26} \times v = 4.59 \times 10^{-20}$ [1 mark]; $v = 4.59 \times 10^{-20}/1.83 \times 10^{-26}$ [1 mark] $v = 2.508 \times 10^{6}$ m/s [1 mark]; 2.51×10^{6} m/s (three significant figures). [1 mark]

Rate of change of momentum

1. a $F = \frac{m\Delta v}{\Delta t} = \frac{60\times15}{0.05}$ [1 mark]; 18 000 N [1 mark]
   b $F = \frac{m\Delta v}{\Delta t} = \frac{60\times15}{0.6}$ [1 mark]; 1500 N [1 mark]
2. The marks are in two bands according to the level of response.

Level 2 (3/4 marks): A clear and coherent explanation is provided which links impact force to how quickly/rate at which the momentum changes.
Level 1 (1/2 marks): Description sufficiently clear to link impact force to impact time.
0: No relevant content.
Indicative content • Impact time is very short if the driver hits his head on the steering wheel/windscreen. • Impact time is much longer if the airbag inflates. • The impact force = $\frac{\text{momentum change}}{\text{time}}$. • Longer impact time reduces the impact force. • If the momentum changes at a slower rate the impact force is smaller.

[4 marks]

Section 6: Waves

Transverse and longitudinal waves

1. a Place a floating object on the water in the middle of the tank [1 mark]. When waves travel across the tank either the cork just moves up and down, or the cork does not move in the direction that the wave is travelling. [1 mark]
   b Right angle/90°. [1 mark]
2. Z is displaced to the right of its undisturbed position [1 mark]; Y is at the centre of a compression [1 mark]; X is at the centre of a rarefaction [1 mark]; W is displaced to the left of its undisturbed position. [1 mark]

Frequency and period

1. Frequency = $\frac{1}{5\times10^{-3}}$ [1 mark]; 200 [1 mark]; Hz [1 mark]
2. a The distance from a point on a wave to the equivalent point on the adjacent wave. [1 mark]
   b $\frac{3}{2.5}$ [1 mark] = 1.2 m [1 mark]
3. $\lambda = 6.0 \div 5$ [1 mark]; 1.2 cm [1 mark]; $T = \frac{1}{(5\div4)}$ [1 mark]; 0.8 s [1 mark]

Wave speed

1. Speed = $\frac{4.0}{2.5}$ = 1.6 m/s [1 mark]; distance = $\frac{v}{f} = \frac{1.6}{2.0}$ [1 mark] = 0.8 m [1 mark]
2. $v = \frac{s}{t} = \frac{2\times80}{(13.8\div30)}$ [1 mark]; 350 m/s [1 mark]; answer to two significant figures as shown. [1 mark]
3. Frequency = 20 ÷ 9.5 [1 mark]; 2.1 Hz [1 mark]; Wavelength = $\frac{42}{4} \times \frac{30}{40}$ [1 mark]; 7.9 cm [1 mark]; Speed = 50 × $\frac{30}{40}$ ÷ 2.2 [1 mark]; 17 cm/s [1 mark]; All three answers given to two significant figures as shown. [1 mark]

Reflection and refraction of waves

1. A: reflection [1 mark]; B: transmission [1 mark]; C: absorption. [1 mark]
2. a The marks are in three bands according to the level of response.

Level 3 (5/6 marks): A detailed and coherent plan covering all steps presented in a logical order. The method would lead to the production of valid results.
Level 2 (3/4 marks): The bulk of the method is described with most of the relevant detail included. It may not be presented in a completely logical sequence.
Level 1 (1/2 marks): Some correct simple statements made but the method would not lead to the production of valid results.
0: No relevant content.

Indicative content

- Position the mirror vertically on a piece of paper.
- Draw along the (front) edge of the mirror then remove from the paper.
- Draw the normal line at 90° to the line corresponding to the front edge of the mirror.
- Use a protractor to draw a line at a specific angle (e.g. 30°) to the normal.
- The specific angle will be the angle of incidence.
- Replace the mirror.
- Use a ray box to direct a ray of light along the drawn line.
- Mark the ray of light reflected from the mirror.
- Remove the mirror and draw in the reflected ray.
- Use a protractor to measure the angle between the reflected ray and the normal.
- Repeat the procedure using a range of angles of incidence.

[6 marks]

b Data agrees with law of reflection within the measurement uncertainty. [1 mark] Either the protractor measures to ±1° or the width of the ray of light adds to the uncertainty in the angle measurements. [1 mark]

3. [1 mark] each for the correctly drawn reflected, refracted and emerging rays, see diagram. Refracted rays must be shown refracted into the block, angle $r < I$, refracted out of the block so that the two rays in air are parallel.

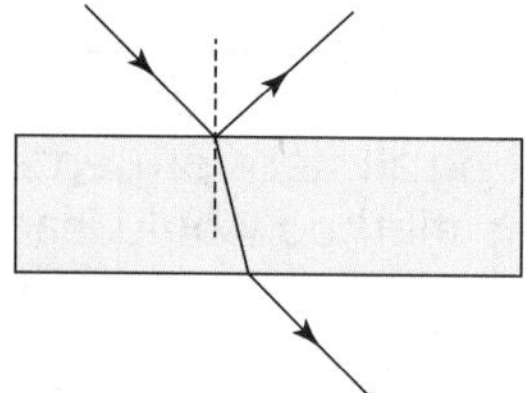

Sound waves

1. a Vibrations are transferred more quickly if molecules are closer together. [1 mark]

b 20 Hz [1 mark] to 20 kHz. [1 mark]

c The energy is absorbed/increased thermal energy store of ear. [1 mark]

2. Wavelength at 20 °C = $\frac{v}{f} = \frac{343}{520}$ [1 mark]; 0.660 m [1 mark]; Wavelength at 0 °C $\frac{v}{f} = \frac{331}{520}$ [1 mark]; 0.637 m [1 mark]; Wavelength change = 0.023 m [1 mark] to two significant figures as shown. [1 mark]

Ultrasound and echo sounding

1. a Depth = $1480 \times \frac{1.92}{2}$ [1 mark]; 1420 m [1 mark]; given to three significant figures as shown. [1 mark]

b Either detecting (shoals of) fish or searching for specific items on the sea bed, e.g. a shipwreck. [1 mark]

2. a Some of the waves are reflected [1 mark] and some are transmitted. [1 mark]

b The time interval between transmitting the pulse and receiving its reflection from the lower side of the block = $\frac{0.20 \times 2}{6100}$ [1 mark] = 6.6×10^{-5} s [1 mark]. If a time interval shorter than 6.6×10^{-5} s is recorded it must be due to the crack. [1 mark]

Seismic waves

1. a [1 mark] for each correct **row** in the table:

Wave	Transverse	Longitudinal	Travels though solids	Travels through liquids
P	no	yes	yes	yes
S	yes	no	yes	no

b S-wave [1 mark]

c P-wave [1 mark]

2. a The S waves are not passing through/transmitted by the core [1 mark]; S-waves don't travel through a liquid [1 mark]; core must be liquid. [1 mark]

b Refraction [1 mark]; the bending occurs when the wave meets a boundary between two different materials where the wave speeds are different. [1 mark]

The electromagnetic spectrum

1. Microwave $\lambda = \frac{3.0 \times 10^8}{2.5 \times 10^9}$ [1 mark]; 0.12 m [1 mark]; number of orders = 6 [1 mark]

2. a Time = $\frac{4 \times 10^{16}}{3.0 \times 10^8} = 1.3 \times 10^8$ s [1 mark]; = $\frac{1.33 \times 10^8}{60 \times 60 \times 24 \times 365}$ [1 marks]; 4 years [1 mark]; given to one significant figure as shown [1 mark]

b Wavelength = $\frac{v}{f} = \frac{3.0 \times 10^8}{2.2 \times 10^{26}}$ [1 mark]; 1.4×10^{-18} m [1 mark]; answer given to two significant figures as shown. [1 mark]

Absorption, transmission, refraction and reflection of electromagnetic waves

1. Visible light [1 mark]
2. The three straight wave fronts drawn refracted (see diagram) [1 mark]; wave fronts in the glass are closer together [1 mark]; arrow shows direction of energy transfer in the glass perpendicular to refracted wave fronts. [1 mark]

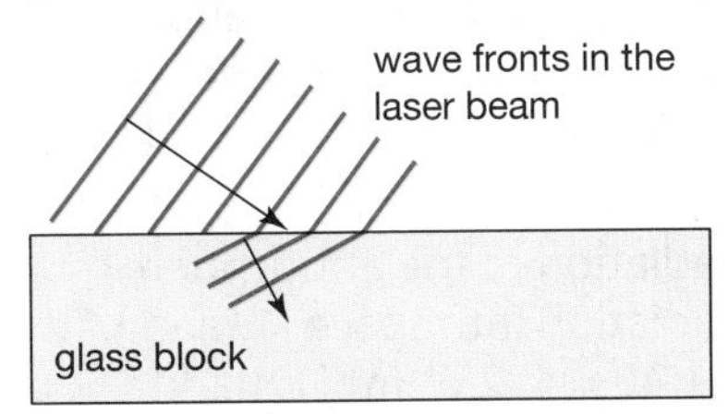

3. [1 mark] for each of the four lines drawn correctly (see diagram below).

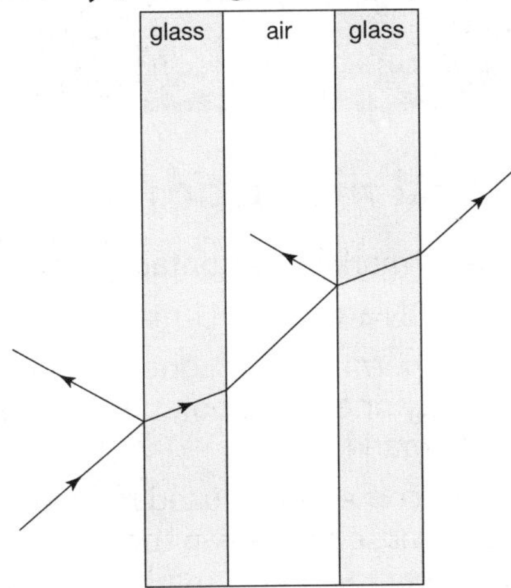

Refracted out of glass into air, two rays in air parallel [1 mark]; refracted into the glass (from air between glass panes), angle $r < i$ [1 mark]; partially reflected ray, angle of incidence = angle of reflection (by eye) [1 mark]; refracted out of glass into air, two rays in air parallel. [1 mark]

4. **a** Decreases. [1 mark]
   b Blue. [1 mark]
   c The shorter the wavelength the greater the amount of refraction. [1 mark]

Emission and absorption of infrared radiation

1. **a** Distance between sensor and surface [1 mark]; temperature of the surface/water inside container. [1 mark]
   b Black surfaces are better emitters than bare metal surfaces. [1 mark] Dull/matt surfaces are better emitters than shiny/polished surfaces. [1 mark]

Uses and hazards of the electromagnetic spectrum

1. A tracer is injected into the body [1 mark]. Gamma rays are very penetrating and can pass through the body to the (gamma) cameras surrounding the body. [1 mark] Gamma rays cause less harm to the inside of the body than alpha or beta radiation. [1 mark]
2. **a** X-rays can penetrate through soft tissue [1 mark]; X-rays cannot penetrate through bone. [1 mark]
   b To ensure that the total radiation dose received by a person does not exceed the safe limit. [1 mark]

Radio waves

1. **a** Oscillations [1 mark] in an electric circuit. [1 mark]
   b The aerial absorbs the radio waves [1 mark]; radio waves induce an alternating current in the aerial [1 mark] of the same frequency as the radio wave. [1 mark]
2. **a** Microwaves can pass through the upper atmosphere/ionosphere [1 mark] because microwaves have a shorter wavelength/ higher frequency than radio waves [1 mark] and so are not reflected. [1 mark]
   b Long wave radio waves have a longer wavelength [1 mark] so are reflected by the upper atmosphere/ionosphere back to Earth. [1 mark]

Colour

1. **a** Specular (reflection) [1 mark]; diffuse (reflection). [1 mark]
   b Specular. [1 mark]
2. **a** White light is made up a spectrum of different colours [1 mark]; the wavelengths in the green part of the visible spectrum are strongly reflected (by the object) [1 mark]; the other wavelengths are (absorbed). [1 mark]
   b Black [1 mark]

3. **a** Both a transparent object and a translucent object allow light to travel through them [1 mark]; it is possible to see through a transparent object but not through a translucent object. [1 mark]
   b The bulb appears red [1 mark]; the filter allows only red light to pass through [1 mark]; all other colours/wavelengths are absorbed (by the filter). [1 mark]

Lenses

1. **a** Convex/converging. [1 mark]
   b The lens brings parallel rays to a focus at the principal focus. [1 mark]
   c The distance from the lens to the principal focus. [1 mark]
   d See diagram below. A ray that continues from the top of the object through the centre of the lens [1 mark]; a ray from the top of the object parallel to the axis, refracted by the lens and continued through F on the right [1 mark]; image shown with an upside down arrow from axis, where rays intersect [1 mark]; all lines must be drawn with a ruler for full marks; no ruler, penalise 1 mark, all lines must be solid not dashed.

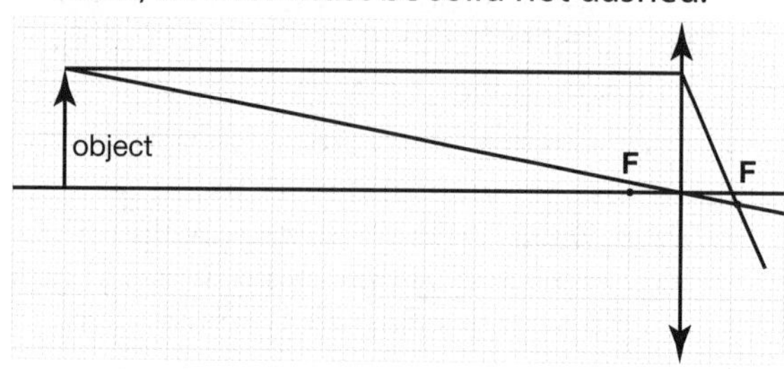

   e [1 mark] each for two of: upside down; real; smaller than the object.
   f Estimate of magnification in the range 0.06–0.12. [1 mark]

A perfect black body

1. **a** An electric current passes through the filament [1 mark] and transfers energy, raising the temperature of the filament [1 mark]; light is emitted by the filament. [1 mark]
   b A bigger electric current passes through the filament [1 mark]. More thermal energy is transferred to the filament [1 mark]. The light is more intense/more energy radiated per second [1 mark]; and its wavelength changes. [1 mark]
2. **a** Absorbs all the light/electromagnetic radiation incident on it [1 mark]; Does not reflect [1 mark] or transmit light/electromagnetic radiation. [1 mark]
   b At a particular temperature [1 mark], a black body would emit the maximum amount of energy possible (for that temperature). [1 mark]

Temperature of the Earth

1. **a** Solar radiation [1 mark]; and radiation from greenhouse gases in the atmosphere [1 mark]. Total: 168 + 324 = 492 (W/m^2). [1 mark]
   b Thermals/latent heat to the atmosphere [1 mark]; radiation to the atmosphere [1 mark]; radiation into space [1 mark]. Total: 102 + 350 + 40 = 492 W/m^2. [1 mark]
   c Constant [1 mark]

Section 7: Magnetism and electromagnetism

Magnets and magnetic forces

1. **a** Repulsive [1 mark]; non-contact. [1 mark]
   b Force = 35 mN attractive. [1 mark]
2. **a** The bar is a permanent magnet if one end is repelled by one of the poles of a known magnet. [1 mark]
   b Each nail becomes an induced magnet [1 mark] because they are in the magnetic field of the bar magnet [1 mark]. The ends of the nails become opposite poles [1 mark]; with a force of attraction between them. [1 mark]

Magnetic fields

1. **a** Correct shape/pattern similar to field of a bar magnet, minimum of four field lines, no field lines crossing [1 mark]; arrows pointing into S [1 mark]; arrows out of N [1 mark]; field lines closest together at the poles. [1 mark]
   b At the geographic North Pole the needle is vertical (with its north pole pointing downwards). [1 mark] At the Equator the needle is horizontal. [1 mark] At the geographic South Pole, the needle is vertical with its north pole pointing in the opposite direction (upwards). [1 mark]

c Seismic waves (created by earthquakes) [1 mark]; S waves are not detected on the opposite side of the Earth from an earthquake [1 mark]; S waves cannot pass through liquids so part of the core must be liquid. [1 mark]

The magnetic effect of an electric current

1. a At least three concentric circles [1 mark]; circles get further apart as they get larger [1 mark]; arrows in clockwise direction. [1 mark]

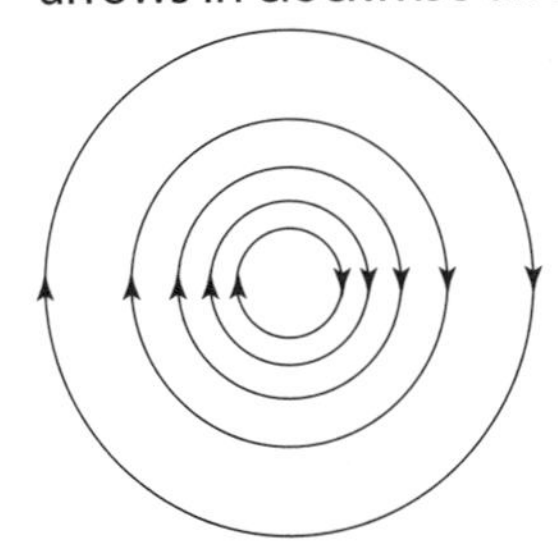

b See diagram below: current direction [1 mark]; N pole [1 mark]; S pole. [1 mark]

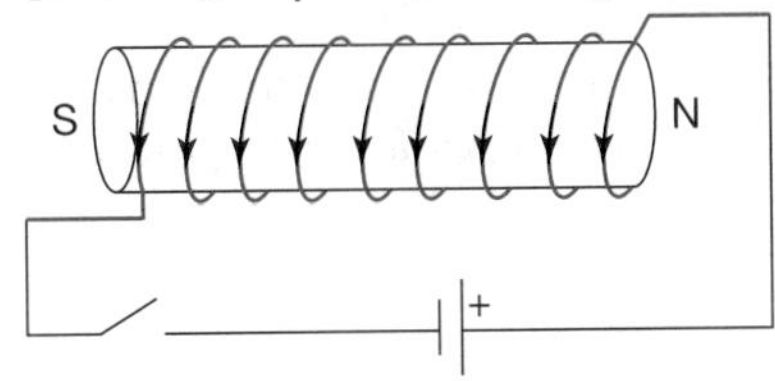

2. The marks are in two bands according to the level of response.

Level 2 (3/4 marks): A detailed and coherent explanation in which each step is logically linked to the previous step.
Level 2: (1/2 marks): Some correct statements but lacks detail and the steps are not logically linked.
0: No relevant content.
Indicative content • Current flows in the solenoid. • The solenoid's current creates a magnetic field. • The magnetic field of the solenoid induces magnetism in the iron strips. • The end of the iron strips in the middle of the solenoid have opposite poles. • The iron strips are attracted together and make contact. • The motor circuit is now complete. • Current flows in the motor. • The motor starts.

[4 marks]

Fleming's left-hand rule

1. a Upwards (out of the field)/towards the top of the page. [1 mark]

b $F = BIl = 300 \times 10^{-3} \times 3.8 \times 0.05$ [1 mark]; 5.7×10^{-2} (N) [1 mark]; 57 mN [1 mark]

2. Weight of the three masses $= mg = 3 \times 3.0 \times 10^{-3} \times 9.8 = 8.82 \times 10^{-2}$ N [1 mark]; $B = \frac{F}{Il}$ [1 mark] $= \frac{8.82\times10^{-2}}{4.1\times0.02}$ [1 mark] $= 1.1$ T [1 mark]; given to two significant figures as shown. [1 mark]

Electric motors

1. a Left side force: up [1 mark]; Right side force: down [1 mark]; Rotation: clockwise. [1 mark]

b $F = BIl = 0.30 \times 0.2 \times 0.04$ [1 mark]; 2.4×10^{-3} N [1 mark]

c Moment $= Fd = 2.4 \times 10^{-3} \times 0.018$ [1 mark]; 4.3×10^{-5} N m [1 mark]; answer given to two significant figures as shown. [1 mark]

d Total moment $= 2 \times 4.3 \times 10^{-5}$ [1 mark]; 8.6×10^{-5} N m [1 mark]

Loudspeakers

1. The marks are in three bands according to the level of response.

Level 3 (5/6 marks): A detailed and coherent explanation covering all the major steps. Each statement is logically linked to the next one.
Level 2: (3/4 marks): Some detail lacking but each statement is logically linked to the next one.
Level 1: (1/2 marks): Some correct statements but lacks detail and the steps are not logically linked.
0: No relevant content.
Indicative content • The alternating potential difference drives an alternating current in the coil. • This creates an electromagnet with an alternating magnetic field. • The field of the permanent magnet interacts with the field of the electromagnet. • The field of the permanent magnet is in opposite directions on opposite sides of the coil.

- The direction of the current in the wire at the bottom of the coil is opposite to the direction of the current in the wire at the top of the coil.
- When the current flows in one direction there is a magnetic force on the turns of the coil which pushes the coil to the right.
- Since the cone is attached to the coil, it also moves to the right.
- When the current reverses direction, there is a magnetic force on the turns of the coil which pushes the coil to the left.
- So the cone moves to the left.
- The repeated to and fro movement of the coil creates pressure variations in air/a sound wave.

[6 marks]

Induced potential

1. a Moving the wire more quickly [1 mark]; moving the wire through a stronger magnetic field. [1 mark]
   b Moving the wire downwards [1 mark]; reversing the direction of the magnetic field. [1 mark]
2. The marks are in two bands according to the level of response.

Level 2 (3/4 marks): A detailed and coherent explanation in which each step is logically linked to the next step.
Level 1 (1/2 marks): Some correct statements but lacks detail and the steps are not logically linked.
Indicative content • Moving the magnet towards the coil induces a potential difference across/in the coil. • The coil is part of a complete circuit so an induced current flows in the coil. • The induced current in the coil creates a magnetic field. • The direction of the induced current opposes the motion of the magnet. • So the top of the coil has to be a north pole to repel the approaching north pole (of the magnet).

[4 marks]

Uses of the generator effect

1. a (Rotating the handle) makes the coil move [1 mark] relative to the magnetic field. [1 mark]
   b One complete cycle shown starting at the origin and ending at 1.0 s [1 mark] (see diagram below); Positive and negative peaks are the same height. [1 mark].

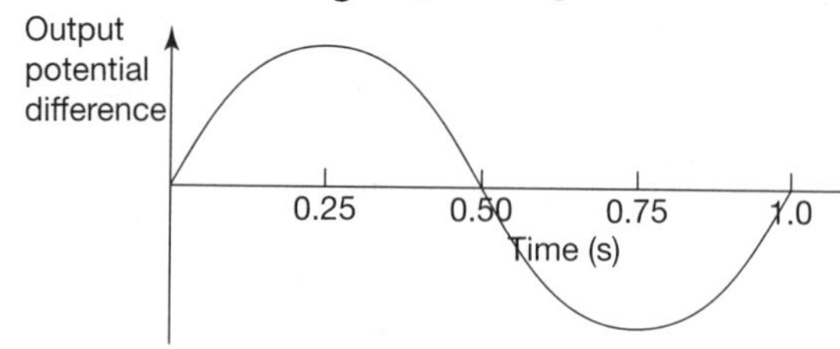

   c Horizontal. [1 mark]
2. One revolution takes 0.04 s [1 mark]; 25 revolutions per second. [1 mark]

Microphones

1. a The vibration of the diaphragm makes the coil move back and forth/vibrate [1 mark]; The coil moves relative to the magnetic field [1 mark] inducing a potential difference. [1 mark]
   b The output potential difference alternates at the same frequency as the sound wave [1 mark].

Transformers

1. a The transformer has to change from 230 V to 15 V/reduce voltage [1 mark]; Diagram shows fewer turns on the secondary coil than the primary so the transformer is a step down/will decrease the voltage. [1 mark]
   b The alternating current in the primary coil [1 mark] creates an alternating magnetic field in the iron core [1 mark]; The alternating magnetic field in the core induces an alternating p.d. in the secondary coil. [1 mark]
   c $\frac{V_p}{V_s} = \frac{n_p}{n_s}$ rearranges to give $n_s = n_p \times \frac{V_s}{V_p}$ [1 mark]; $500 \times \frac{15}{230}$ [1 mark]; 33 [1 mark]; given to the nearest whole number as shown. [1 mark]
   d Rearrange $V_s I_s = V_p I_p$ to give $I_p = \frac{V_s I_s}{V_p}$ [1 mark]; $\frac{15.0 \times 8.16}{230}$ [1 mark]; 0.532 A [1 mark]; answer given to three significant figures as shown. [1 mark]

Section 8: Space physics

Our Solar System

1. Gravity/gravitational force [1 mark]; attractive [1 mark]; non-contact. [1 mark]
2. Five. [1 mark]

The life cycle of a star

1. The marks are in two bands according to the level of response.

Level 2 (3/4 marks): A clear, coherent explanation using appropriate scientific terms.
Level 1 (1/2 marks): Some detail may be missing, or the response may lack a logical structure.
0: No relevant content.
Indicative content • The core of a star is at a very high temperature. • Allowing hydrogen nuclei to undergo fusion. • Helium nuclei are formed. • Later in the star's life all the hydrogen (in the core) has been fused/used up • Helium nuclei can undergo fusion to produce larger (heavier) elements (up to iron). • The more massive the star, the hotter its core, so heavier elements (up to iron) are produced.

[4 marks]

2. a The greater the mass (of a star) then the shorter the time it spends as a main sequence star. [1 mark]
   b [1 mark] for an answer in the range 3–5 billion years.
3. The marks are in three bands according to the level of response.

Level 3 (5/6 marks): A clear, detailed coherent description including specialist terms presented in the correct sequence.
Level 2 (3/4 marks): Major changes presented clearly in the correct sequence.
Level 1 (1/2 marks): Some relevant information but some detail missing and may not be in the correct sequence.
0: No relevant content.
Indicative content • The star expands and becomes a red giant. • It continues to produce energy by nuclear fusion. • Fusion produces carbon atoms. • The outer layers (of the star) are blown away. • Leaving its smaller, much denser core. • This collapses/compresses due to own gravity. • The temperature rises and... • ... the star glows much brighter. • It is now described as a white dwarf. • Cannot generate further energy by nuclear fusion. • Cools and dims as time passes. • It is described as a black dwarf when the star stops emitting energy/radiation.

[6 marks]

Orbital motion, natural and artificial satellites

1. a Each planet is travelling at a steady speed round the Sun [1 mark] but its direction of motion is constantly changing so the velocity is too. [1 mark]
   b Time of orbit = $365 \times 24 \times 60 \times 60 = 3.154 \times 10^7$ (s) [1 mark]; speed = $\frac{9.40 \times 10^{11}}{3.154 \times 10^7}$ [1 mark]; $2.98(03) \times 10^4$ m/s [1 mark]; 2.98×10^4 m/s (three significant figures). [1 mark]
2. a To maintain a stable orbit at a specific radius [1 mark] the speed must have a particular value (if speed not correct, radius of orbit will change). [1 mark]
   b Any two of the following for [1 mark] each:
   MetOp-B's orbit is much closer to the Earth than the orbit of Inmarsat-4, which makes it better for making observations of clouds/rainfall/temperatures/wind speeds; MetOp-B's orbit enables it to make many orbits each day so it can detect changes in the weather; MetOp-B's orbit enables it to observe developing weather patterns over the whole of the Earth rather than just a specific area/different geographical view with each orbit.

Red-shift

1. **a** The wavelength of the light is increased [1 mark] because the galaxy is moving away (from the Earth). [1 mark]

   b The longer the distance the faster the (recession) speed. [1 mark]

2. **a** 26 000–27 000 km/s. [1 mark]

   b 150 Mpc. [1 mark]

3. The marks are in two bands according to the level of response.

Level 2 (3/4 marks): A clear, coherent description making the link between the observed red-shift and the Big Bang theory.
Level 1 (1/2 marks): Some relevant information but no clear link between observed red-shift and the Big Bang theory.
0: No relevant content.
Indicative content • The red-shift shows that distant galaxies are moving away (receding). • The further away the galaxy the bigger the red-shift. • More distant galaxies are moving away faster. • Evidence that the Universe is expanding and the start of the Universe could have been from a single explosion. • The Big Bang theory proposes that the Universe began in a small space that was hot and dense and then expanded. • (Massive) explosion/rapid expansion sent all matter outwards/caused Universe to expand.

[4 marks]

Dark matter and dark energy

1. Matter (or mass) that cannot be directly observed [1 mark] but makes up most of the mass in the Universe. [1 mark]

2. **a** The light from the distant object has been travelling for a very long time/billions of years [1 mark]; the light detected was emitted from the object a very long time ago/ billions of years ago. [1 mark]

   b Dark energy [1 mark] opposes gravity/force/ field/energy pushing galaxies apart. [1 mark]

Notes

Notes

Notes

Notes